Titanium

Industrial Base, Price Trends, and Technology Initiatives

Somi Seong, Obaid Younossi, Benjamin W. Goldsmith

With Thomas Lang, Michael Neumann

Prepared for the United States Air Force

Approved for public release; distribution unlimited

The research described in this report was sponsored by the United States Air Force under Contract FA7014-06-C-0001. Further information may be obtained from the Strategic Planning Division, Directorate of Plans, Hq USAF.

Library of Congress Cataloging-in-Publication Data is available for this publication.

ISBN: 978-0-8330-4575-1

Published 2009 by the RAND Corporation
1776 Main Street, P.O. Box 2138, Santa Monica, CA 90407-2138
1200 South Hayes Street, Arlington, VA 22202-5050
4570 Fifth Avenue, Suite 600, Pittsburgh, PA 15213-2665
RAND URL: http://www.rand.org/
To order RAND documents or to obtain additional information, contact
Distribution Services: Telephone: (310) 451-7002;
Fax: (310) 451-6915; Email: order@rand.org

Preface

Titanium is an important raw material that accounts for a significant portion of the structural weight of many military airframes. It offers an excellent set of properties, such as high strength-to-weight ratio, high strength at high temperatures, corrosion resistance, and thermal stability, that make it ideal for airframe structures. However, in recent years a combination of multiple factors caused a major spike in titanium prices that is expected to significantly influence the acquisition costs of future military aircraft.

This monograph examines the titanium industrial base, production technology, and demand characteristics important to the price of military aircraft. In particular, it addresses the factors underlying price fluctuations in the titanium market in an effort to better forecast economic risks involved in the market and to improve estimates of the future cost of military airframes. We attempt to identify what triggered the unprecedented dramatic increase in titanium metal prices between 2003 and 2006 by presenting an analysis of the raw material markets themselves. The monograph also reviews new titanium manufacturing techniques and assesses their implications for the production cost of future military airframes. In addition, it analyzes both supply- and demand-side determinants of prices and their future prospects.

The research reported here was sponsored by then–Lt Gen Donald J. Hoffman when he was the Military Deputy, Office of the Assistant Secretary of the Air Force (Acquisition), SAF/AQ, and Blaise Durante, SAF/AQX, and was conducted within the Resource Management Program of RAND Project AIR FORCE (PAF). The project's technical monitor is Jay Jordan, Technical Director of the Air Force Cost Analysis Agency.

This monograph should interest those involved with the acquisition of systems for the Department of Defense and those involved in the field of cost estimation, especially for titanium-intensive systems.

This document is one of a series from a PAF project entitled "Weapon System Costing Umbrella Project." The purpose of the project is to improve the tools used to estimate the costs of future weapon systems. It focuses on how recent technical, management, and government policy changes affect cost. Another PAF report that addresses military aircraft material cost issues is *Military Airframe Costs: The Effects of Advanced Materials and Manufacturing Processes,* MR-1370-AF, 2001, by Obaid Younossi, Michael Kennedy, and John C. Graser, which examines cost-estimating methodologies and focuses on military airframe materials and manufacturing processes. This report provides cost estimators with factors useful for adjusting and creating estimates based on parametric cost-estimating methods.

RAND Project AIR FORCE

RAND Project AIR FORCE (PAF), a division of the RAND Corporation, is the U.S. Air Force's federally funded research and development center for studies and analyses. PAF provides the Air Force with independent analyses of policy alternatives affecting the development, employment, combat readiness, and support of current and future aerospace forces. Research is conducted in four programs: Aerospace Force Development; Manpower, Personnel, and Training; Resource Management; and Strategy and Doctrine.

Additional information about PAF is available on our Web site:
http://www.rand.org/paf

Contents

Figures

Tables

Summary

Titanium is an important raw material accounting for a significant portion of the structural weight of most military airframes. It offers an excellent set of properties, such as high strength-to-weight ratio, high strength at high temperatures, corrosion resistance, and thermal stability, that make it ideal for airframe structures. However, in recent years a combination of factors caused a major spike in titanium prices that is expected to influence the acquisition costs of future military aircraft.

Between 2003 and 2006, the price of this expensive metal increased at an unprecedented rate, more than doubling during this period. Government and industry observers said this was the first time a global materials supply concern has affected the defense sector since the steel shortage after World War II (Murphy, 2006). They also noted that the short supply of titanium might influence delivery schedules for military aircraft and weapons (Toensmeier, 2006). There are worries that titanium shortages may substantially raise the program cost of the F-35 (Murphy, 2006; *Defense Industry Daily,* 2006). Although prices of titanium products have fluctuated over the years, the recent price surge was extreme compared to previous fluctuations (see Figure S.1).

Study Objective and Approach

The Office of the Assistant Secretary of the Air Force for Acquisition asked PAF to conduct this study in order to better understand the factors underlying price fluctuations in the titanium metals market, to better forecast economic risks involved in the market, and to improve

Figure S.1
Producer Price Index Trend for Titanium Mill Products, 1971–2006

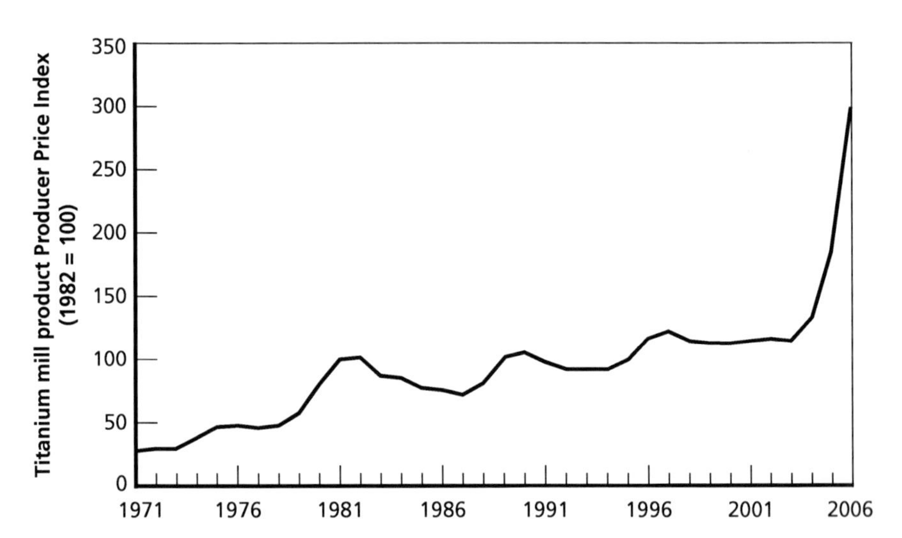

SOURCE: Producer Price Index data from the Bureau of Labor Statistics.
RAND *MG789-S.1*

estimates of the future cost of military airframes. To do so, we attempted to answer three primary questions:

- What triggered the recent titanium price surge?
- What are the future market prospects and emerging technologies?
- What are the policy implications for U.S. military buyers of airframe structures and other titanium-intensive weapons?

Although a previous RAND study (Younossi, Kennedy, and Graser, 2001) focused on the costs of processing raw materials into airframe parts, this study analyzes the actual raw material markets. It also reviews new manufacturing techniques and assesses their implications for the production cost of future military airframes.

Based on literature reviews and analyses of historical data available in defense and commercial industries, this monograph assesses the past trends, current changes, and future prospects for each of the titanium price determinants and their relative importance. In particular, the study analyzes both supply- and demand-side price determinants and their future prospects.

To widen our understanding of the titanium industry, we conducted interviews with experts from the titanium manufacturing and processing industries, the aircraft manufacturing industry, and government agencies compiling titanium price data, such as the United States Geological Survey and the Bureau of Labor Statistics.[1] (See pp. 1–6.)

Titanium Is Expensive to Produce

Titanium is expensive to refine, process, and fabricate. In terms of processing cost per cubic inch, titanium is five times more expensive than aluminum to refine and more than ten times as expensive as aluminum to form into ingots and to fabricate into finished products. Titanium sponge is the commercially pure form of titanium metal that is refined from titanium ore.[2] Titanium ingot is produced from titanium sponge, titanium scrap, or a combination of both.[3] Titanium mill products, such as plate, sheet, billet, and bar, are produced from titanium ingot through such primary fabrication processes as rolling and forging. Titanium parts are then produced from mill products by means of secondary fabrication processes, such as forging, extrusion, hot and cold forming, machining, and casting. Fabrication is the most costly processing stage, followed by sponge production. (See pp. 7–20.)

What Triggered the Recent Titanium Price Surge?

It is a common belief that cyclical fluctuations of titanium prices are mainly driven by demand-side events, especially aircraft demand cycles. However, the Producer Price Index (PPI) for titanium mill shapes in the United States was relatively insensitive to the declining demand from

---

[1] Appendix B contains questions asked of industry experts.

[2] It is called *sponge* because of its porous, sponge-like appearance.

[3] Titanium scrap is a by-product of the fabrication processes.

the commercial aircraft industry during the previous downturn (1998–2003), contrary to common belief. This is because world titanium demand did not decrease as severely as *commercial* aerospace demand. In the global titanium market, industrial demand, historically more stable than aerospace demand, had dominated aerospace demand since the mid 1990s. The industrial titanium market bottomed out in 2001, earlier than the aerospace market, which hit bottom in 2003. Driven by the growth in industrial demand, global titanium demand was already at its previous peak level in 2004. This contributed to amplifying the impact on titanium price and supply availability of the historic aircraft order surge in 2005 and 2006. Given that industrial demand dominates the global market, commercial aerospace demand is not the only major driver of titanium market prices. In fact, the extreme price volatility in the recent titanium market resulted from the coincidence of various supply-side and demand-side price drivers.

Supply-Side Drivers

On the supply side, prices of titanium sponge and scrap began increasing sharply even before the significant surge in commercial aircraft orders in 2005 and 2006. There was an extreme shortage of titanium scrap in 2003, because of the low aircraft production rate, which resulted in less recycled scrap. This coincided with the period during which China's dramatic growth in steel consumption[4] drove up the prices of ferrotitanium,[5] an alloy used in the steel production process. The ferrotitanium price surge led to increased demand for titanium scrap and sponge, both of which are close substitutes for ferrotitanium in steel production. The cross-market substitution effect was significant, because the steel market size was 10,000 times that of titanium. In addition, the Defense Logistics Agency titanium sponge stockpile depletion in 2005 also coincided with the sponge and scrap market shortage, worsening the titanium raw material supply shortage. The

[4] World steel prices increased dramatically in 2003–2004, driven mainly by China's strong demand and world economic recovery.

[5] Ferrotitanium is used in steel production processes for deoxidation, achieving a finer grain structure, and controlling carbon and nitrogen.

stockpile depletion, which had been authorized by Congress, started in 1997; by 2005, there was no titanium sponge left in the stockpile. Since the supply of titanium raw materials was already tight in 2003–2005, the additional demand shock from the record-high level of commercial aircraft orders in 2005 and 2006 intensified the shortage. In addition, titanium metal suppliers were not able to respond quickly to ameliorate the supply shortage. In particular, expanded sponge capacity required building an additional factory, which would take about three years and an investment of $300 million to $400 million. Right before the recent demand surge, titanium producers had suffered from several lean years, and some producers were on the verge of bankruptcy. As a result, the producers hesitated to invest in capacity expansion until they were assured that increased demand would continue for at least the next several years. (See pp. 35–50.)

Demand-Side Drivers

On the demand side, there have been three main demand drivers in the aircraft manufacturing industry in recent years. First, commercial aircraft orders skyrocketed as both Boeing and Airbus received record levels of orders during 2005 and 2006.[6] Second, the average level of titanium content per aircraft rose significantly, which meant that increases in aircraft orders in turn amplified the demand for titanium. Third, the demand for titanium in military aircraft production also increased significantly, as full-time production of the F-22A Raptor began in 2003.[7] These three demands coincided to create a record-breaking increase in titanium demand.

In addition, increases in military armor and industrial demand for titanium added to the demand surge from the aircraft industry. Even before the surge in aircraft demand, the global titanium market

6 We obtained commercial aircraft order and delivery data from the Boeing and Airbus Web sites.

7 Full production of the F-22 was funded in fiscal year (FY) 2003. Refer to the Defense Acquisition Management Information Retrieval Web site, Selected Acquisition Report for the F-22A, December 31, 2006.

was already tight because of high demand from the industrial equipment industry, the steel industry, and other titanium users.

Titanium price volatility was further exacerbated by an increase in spot transactions on the titanium market in 2005 and 2006. During this period of demand surge, even aircraft manufacturers, which normally rely on long-term contracts for their titanium, had to procure titanium on the spot market because of the supply shortage and long lead times.[8] In such a strong seller's market, titanium prices were subject to the titanium producer's bargaining power.

On the whole, increased demand for titanium exceeded the available supply of scrap and sponge, as well as the production capacity for new titanium metal. Given the fact that titanium sponge production capacity expansion requires a high capital investment and long lead times, sponge supply expansion was simply not responsive enough to meet the unexpected surge in demand over the short run. Moreover, given the long record of excess capacity in the industry, titanium producers were reluctant to invest in capacity expansion until they were assured that the strong demand was not temporary. The market imbalance was further worsened by the spurt of speculative purchasing on the spot market, which amplified price volatility. Titanium prices skyrocketed and remained extremely volatile from 2003 to 2006. (See pp. 51–71.)

Market Prospects and Emerging Technologies

Titanium Markets in the Near Future

By the end of this decade, the world titanium sponge production capacity is expected to almost double its 2005 capacity, growing to approximately 217,970 tons per year. In response to the recent demand surge, many titanium metal producers have announced increases in titanium sponge capacity or have taken steps to increase in the near future. If

---

[8] Some of the military aircraft contractors had minimal protection of long-term contracts and were exposed to the risk of price volatility and supply shortage to a greater extent, as they had to purchase titanium for one lot production at a time.

new titanium sponge plants become fully operational as planned, Japan and China will be the top titanium sponge producers in the world, followed by Russia and the United States.

For market prospects, we examined three potential scenarios of world titanium demand: optimistic, base, and pessimistic. In each scenario, we assumed a certain combination of annual average growth rates in titanium demand from the aerospace and industrial market segments and then calculated the projected demand in 2010 in relation to the actual 2005 demand.

We do not attach probabilities to each of the potential future scenarios; rather, we use the scenarios to bound predictions for the future. As a result, different combinations of demand and supply scenarios will result either in a variety of potential market imbalances or in market equilibrium.

Assumptions regarding the following three factors heavily influence the future titanium market outlook:

1. realization of the capacity expansion plans by titanium suppliers, including American and Chinese producers
2. the Boeing 787 build rate and demand from other titanium-intensive aircraft
3. continued Chinese economic growth and Chinese consumption of steel, titanium, and other metals that are related to world titanium demand and supply conditions.

By comparing the demand scenarios and production capacity expansion plans, we determined that the titanium industry's current capacity expansion plans appear to be based on the future demand expectations inherent in the optimistic scenario. Therefore, if the base demand scenario is realized (instead of the optimistic scenario) and the world titanium production capacity expands as planned, we expect there will be excess production capacity in the titanium market by 2010. (See pp. 73–82.)

Emerging Technologies

Breakthroughs. Of the experts we interviewed, only a few were optimistic about any dramatic changes in titanium metal extraction, processing, and production technologies that may be realized within the next ten years. In addition, the titanium industry has not identified any particular technology that is worthy of an aggressive investment for a medium-term (three- to five-year) return. Titanium companies are taking a "wait and see" position on significant technological breakthroughs.

Technologies with Cost-Saving Potential. After reviewing the literature and conducting discussions with industry experts, we developed a list of emerging technologies with at least marginal cost-saving potential. These technologies are classified into five categories:

- improved extraction and refinement
- powder metallurgy
- single-melt refining
- solid freeform fabrication
- improved machining.

We then assessed these technologies based on their time frame, feasibility of application to the market, and potential for cost savings.

The greatest potential cost savings lie in the combination of improved extraction processes and powder metallurgy, a specialized titanium production process that would limit waste and remove many steps from the current production process. If successful, the higher yield and increased speed of this process would expand the amount of titanium on the market and considerably shorten lead times, dramatically changing the titanium industry over the next decade. However, this combination of developments is unlikely to occur in the near or mid term and is still technically uncertain.

Single-melt refining (instead of multiple-melt refining) and improved machining also would improve production yields and save time and energy. Although the savings from these improvements are expected to be smaller than the savings offered by improved extraction and powder metallurgy, they are also expected to be much steadier and more consistent.

Across these new technologies, most savings will be realized by improved yields resulting from reduced waste during processing and part fabrication. Improved labor efficiency will yield some savings, especially during the fabrication process. Energy savings should be an important, but much smaller, proportion of the savings, primarily concentrated in improvements during initial extraction and melting.

The emerging technologies have the potential to reduce costs sufficiently to open new markets, such as military ground vehicles. However, it will take a long time for these technologies to influence the cost of aerospace-grade titanium substantially.

Barriers to Adopting New Technologies. A major barrier to adoption of new technologies in aerospace applications is the required certification of new materials. Aerospace manufacturing standards are typically based either on judgments by a government body, such as the Federal Aviation Administration or the U.S. Air Force, or on standards set by the primary aircraft manufacturers. Within the Air Force, materials and processes must be certified separately for each program. The certification process typically lasts 18 to 24 months and requires extensive qualification processes. In the course of this process, a company must manufacture test articles and validate their properties at its own expense. The cost of this process prevents companies from attempting to certify materials until they are quite certain of their performance and properties. Consequently, an innovative titanium product or process must be used for several years in other applications before designers will consider it for aerospace uses. (See pp. 82–94.)

Policy Implications

Based on the findings of this study, we suggest policy measures in five areas: improving contract practices, monitoring market trends, reducing buy-to-fly ratios, optimizing scrap recycling, and exploring new technological opportunities. (See pp. 95–107.)

Acknowledgements

We thank the sponsors of this project, then–Lt Gen Donald J. Hoffman, former SAF/AQ, and Blaise Durante, SAF/AQX. We are grateful to Mr. Durante for his long-term support of PAF's cost analysis and acquisition research. We extend our appreciation to Maj Gen Jeffrey Riemer, former F-22A Program Executive Officer, for allowing us to accompany him on his many contractor visits. We also thank the project monitors, Richard Hartley, Deputy Assistant Secretary for Cost and Economics, Office of the Assistant Secretary of the Air Force for Financial Management and Comptroller, and Jay Jordan, Technical Director for the Deputy Assistant Secretary for Cost and Economics. We are grateful to Edward Rosenberg of the F-22A Program Executive Office for helping us establish contacts with many titanium vendors and experts.

This study benefited greatly from discussions with experts from the titanium manufacturing and processing industries and the aircraft manufacturing industry, as well as government agencies that compile titanium price data. The authors are thankful for their valuable insights.

We would like to acknowledge the following principal points of contact at each organization we visited or interviewed with. At government organizations, we thank Jane Adams, Army Research Laboratory; Joseph Kowal, Bureau of Labor Statistics; Leo Christodolou, Defense Advanced Research Projects Agency; and Joseph Gambogi, United States Geological Survey.

Among industry organizations, we thank Thomas Bayha, Allvac Incorporated; Thomas Blanchard, Christopher DeForest, Jeffrey K.

Hanley, Barton Moenster, and Kevin Slattery, The Boeing Company; Oscar Yu, RTI International; Henry Seiner and David Tripp, Titanium Metals Corporation; Kevin Lynch, Wyman-Gordon; and Michael T. Hyzny, DuPont Titanium Technologies.

The thoughtful input from our reviewers, Jan Miller, Steven W. Popper, Laura Baldwin, and Cynthia Cook, did much to improve the manuscript. Finally, we thank Brian Grady, Megan McKeever, and Regina Sandberg for their research and administrative support and Miriam Polon for editing the monograph.

Abbreviations

AIA	Aerospace Industries Association of America
AIAA	American Institute of Aeronautics and Astronautics
AISI	American Iron and Steel Institute
AMPTIAC	Advanced Materials and Processes Technology Information Analysis Center
ATI	Allegheny Technologies Incorporated
BLS	Bureau of Labor Statistics
BMI	bismaleimide
BTF	buy-to-fly ratio
CAGR	compounded annual average growth rate
CFRP	carbon fiber reinforced polymer
CIP	cold isostatic pressing
DLA	Defense Logistics Agency
DNSC	Defense National Stockpile Center
DoD	Department of Defense
EBM	electron beam melting
FY	fiscal year
HIP	hot isostatic pressing
HSM	high-speed machining
IISI	International Iron and Steel Institute
ITA	International Titanium Association

LTA	long-term agreement
MBW	material buy weight
MER	MER Corporation
MFW	material fly weight
MIM	metal injection molding
ODUSD-IP	Office of the Deputy Under Secretary of Defense for Industrial Policy
PAF	Project AIR FORCE
PAM	plasma arc melting
P/M	powder metallurgy
PPI	Producer Price Index
PRC	People's Republic of China
PREP	preparation
RMI	RTI International Metals, Inc.
SAR	Selected Acquisition Report
TIMET	Titanium Metals Corporation
TiO2	titanium dioxide
Ti-6AL-4V	titanium alloyed with 6 percent aluminum and 4 percent vanadium
USGS	United States Geological Survey
VAR	vacuum arc remelting
VSMPO	Verkhnaya Salda Metallurgical Production Association

CHAPTER ONE

Introduction

Background

Titanium is an important metal, accounting for a significant portion of the structural weight of many military airframes. It offers an excellent set of properties, such as a high strength-to-weight ratio, corrosion resistance, and thermal stability, that make it ideal for airframe structures. For example, titanium contributes about 39 percent of the structural weight of the F-22A Raptor (Phelps, 2006). Similarly, a legacy air superiority fighter such as the F-15 includes approximately 32 percent titanium in its structural weight. The Navy's F/A-18 E/F includes about 21 percent titanium in its airframe structure (Younossi, Kennedy, and Graser, 2001). Figure 1.1 displays a time trend in the use of titanium in military aircraft.

Although titanium constitutes a relatively significant percentage of the aircraft's structural weight as measured by material fly weight (MFW), the amount of titanium material necessary to produce each plane, called the titanium material buy weight (MBW), is many times more than the amount actually included in the finished aircraft. Because of the reactive properties of the metal and the multistep refinement, machining, and fabricating processes, a significant amount of titanium scrap is generated during the airframe production process. The ratio of the total weight of purchased raw material to the weight of the finished part included in the airframe is commonly referred to as the buy-to-fly (BTF) ratio.

For example, the titanium MBW is more than ten times the MFW for the F-22A Raptor; each F-22A requires about 50 metric tons

Figure 1.1
Percentage of Titanium in the Structural Weight of Selected Military Aircraft

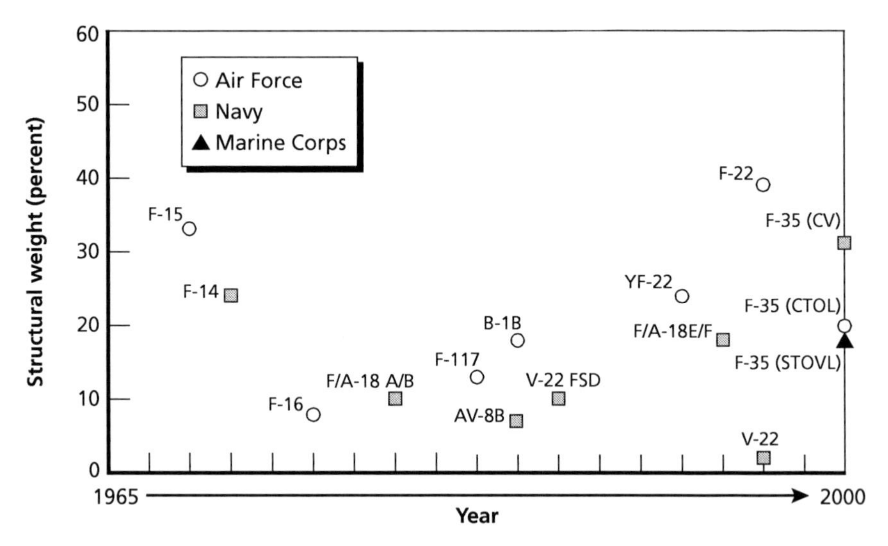

SOURCE: Younossi, Kennedy, and Graser, 2001, updated with recent F-22 and F-35 information.
RAND *MG789-1.1*

of titanium.[1] By comparison, the BTF ratio for an F-15 is about 8, and each plane requires about 30 metric tons of titanium.[2] If we assume the BTF ratio of the F-35 is similar to that of the F-22A, then the F-35 will require about 15 tons of titanium per plane on average.[3]

Although the raw material cost of titanium accounts for a relatively small portion of the unit recurring flyaway cost, a sharp increase in the titanium metal price will influence the acquisition cost of future

1 Titanium MBW and MFW statistics were obtained from the F-22 Program Office in November 2007.

2 The F-15 BTF calculation is based on the estimates of MBW and MFW in Schmitt, 1993.

3 Because the F-35 configuration is not yet mature, the average BTF ratio estimates vary widely from 7 to 26, depending on sources and the time of estimation. The sources include F-35 Joint Program Office, Lockheed Martin Aeronautics Company, and ODUSD-IP, 2005. The average BTF ratio means the weighted average BTF ratios of the three types of the F-35: F-35 CTOL, F-35 STOVL, and F-35 CV.

military aircraft.[4] According to the Office of the Deputy Under Secretary of Defense for Industrial Policy (ODUSD-IP, 2005), a 50 percent increase in titanium prices would increase the unit price of the F-22A by $1.3 million, which is about 1 percent of the plane's unit recurring flyaway cost. However, titanium prices almost tripled between 2003 and 2006, which means the unit recurring flyaway cost of an F-22 might have increased about 6 percent.

Recently, the price of this expensive metal has increased at an unprecedented rate. The Producer Price Index (PPI) for titanium mill shapes more than doubled in three years, from 114 in 2003 to 300 in 2006.[5] During the same period, the average sales price of mill products[6] by major titanium metal producers—those who receive a significant portion of their sales from long-term contracts[7]—also nearly doubled during this time frame.[8]

Government and industry observers say this is the first time that a global materials supply concern has affected the defense sector since the steel shortage that followed World War II (Murphy, 2006). They also note that the titanium supply shortage may influence delivery schedules for military aircraft and weapons (Toensmeier, 2006). There are worries that titanium shortages may substantially raise the program cost of the F-35, previously called the Joint Strike Fighter (Murphy, 2006; *Defense Industry Daily,* 2006).

---

[4] Unit recurring flyaway cost is the cost of the airframe, propulsion, armament, electronic fire control, and similar air-vehicle items. Airframe is usually the most significant cost element.

[5] The PPI tracks the average change in net transaction prices that domestic producers and service providers receive for the products and services that they make and sell. Since the PPI tracks transaction prices, it is based on both spot market prices and long-term contract prices, similar to the universe of transactions in the producer market place. PPI statistics were downloaded from the Bureau of Labor Statistics Web site.

[6] Mill products such as billet, bar, plate, sheet, tube, and wire are fabricated through forging or rolling processes.

[7] In the aerospace industry, long-term contracts are often referred to as long term agreements, or LTAs.

[8] According to the Titanium Metals Corporation Annual Report (TIMET, 2006), mill product average price increased from $31.50 in 2003 to $57.85 in 2006.

Prices of titanium mill products have fluctuated cyclically over the years. However, as shown in Figure 1.2, the recent price surge was extreme compared to previous fluctuations. What caused the unprecedented price increase in titanium metal products? What are the implications for the future cost of military airframes? This monograph will explore these questions further.

Study Objective

This study aims to understand the factors underlying price fluctuations in the titanium metals market in order to better forecast the economic risks involved in the market and to improve estimates of the future costs of military airframes. It attempts to answer the question of what triggered the unprecedented dramatic increase in titanium metal prices between 2003 and 2006. While a previous RAND study (Younossi, Kennedy, and Graser, 2001) focused on the costs of processing raw materials into airframe parts, this study analyzes the actual raw mate-

Figure 1.2
Producer Price Index Trend for Titanium Mill Products, 1971–2006

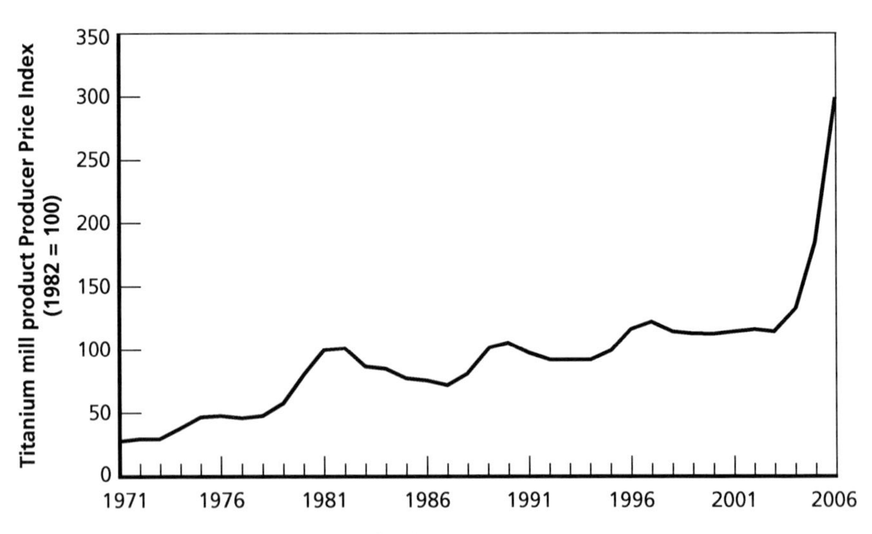

SOURCE: PPI data from the Bureau of Labor Statistics.
RAND *MG789-1.2*

rial markets. It also reviews new manufacturing techniques and assesses their implications for the production cost of future military airframes.

Approach

Based on literature reviews and analysis of historical data available in defense and commercial industries, we assess past trends, current changes, and future prospects for each of the titanium price determinants and their relative importance.

In particular, we analyze both supply- and demand-side price determinants for titanium. Some of these price determinants are detailed in Table 1.1.

To confirm our understanding of the industry, we conducted interviews with experts from the titanium manufacturing and processing industries and the aircraft manufacturing industry,[9] as well as government agencies that compile titanium price data, such as the United

Table 1.1
Price Determinants of Titanium

Supply-Side Determinants	Demand-Side Determinants
Major suppliers	Major buyers
Degree of competition among suppliers such as number of suppliers, entry, and exit	Downstream industries
Distribution of production capacity over suppliers	Market conditions of downstream industries and their influence on demand
Geographic distribution of industrial base (U.S., China, etc.)	Market for titanium substitutes
Cost-reducing technological changes in the industry	Relative importance of the U.S. military buyer in the titanium market
Industrywide learning curve, if relevant	Foreign demand trends and their impact on prices for U.S. military buyers
Ore and other raw materials costs	Other factors that may influence the demand-side market conditions
Other factors that may influence cost to suppliers	

[9] Typical questions posed to industry experts are illustrated in Appendix B.

States Geological Survey (USGS) and the Bureau of Labor Statistics (BLS).

Outline of the Monograph

Chapter Two presents the basic characteristics of titanium metal, the products involved, production processes, and the production cost structure. Chapter Three discusses the titanium industrial base and other market characteristics including major suppliers, distribution of production capacity over suppliers and geographic regions, major consumers of titanium, and titanium price trends. Chapter Four examines how the supply-side drivers of titanium price fluctuations unfolded to create the recent turmoil in the titanium market. Chapter Five analyzes demand-side drivers of titanium price fluctuations. Chapter Six reviews future prospects for the titanium market and discusses cost-saving technology trends. Chapter Seven derives policy implications for U.S. military buyers of airframe structures and other titanium-intensive weapons.

CHAPTER TWO

Titanium Processing

This chapter provides basic information on titanium, its properties, products, and processing techniques. It concludes with a discussion of titanium-processing cost drivers.

Titanium and Its Properties

Titanium's many useful properties make it a critical material in building aerospace systems. Titanium has a high strength-to-weight ratio, corrosion resistance, and thermal stability. It is as strong as steel but 45 percent lighter. It is approximately 60 percent heavier than aluminum but is more than twice as strong as the most commonly used aluminum alloy (Barksdale, 1968). Its resistance to corrosion is significantly higher than that of stainless steel. In addition, titanium's coefficient of thermal expansion is significantly less than that of ferrous alloys, copper-nickel alloys, brass, and many stainless steels.

However, titanium's main drawback is its high price—titanium metal is more than five times as expensive as aluminum. This is not because titanium ore is scarce.[1] In fact, titanium is the fourth-most abundant metal in the earth's crust and the ninth-most common element on the entire planet (Kraft, 2004; Gerdemann, 2001;

[1] According to TIMET, 2006, the availability of rutile ore (the major titanium-containing ore) will not be a problem in the foreseeable future.

Cariola, 1999). However, titanium is expensive to refine, process, and fabricate.[2]

Titanium Metal Products

The titanium industry produces a variety of products—titanium sponge, ingot, and mill products. These mainly intermediate goods are produced when titanium ore is refined, melted, and fabricated into a metal.

Ores and Concentrates

Most of the titanium ore processed in the United States comes from either Australia or South Africa. Titanium is found in both rutile and ilmenite (iron titanium oxide) ores, which contain about 95 percent and 70 percent titanium, respectively.[3] All titanium metal production begins with rutile (titanium oxide, or TiO2). High-titania slag, produced by ilmenite smelting, is the first, most important step in the production of synthetic rutile. More than 80 percent of titanium resources come from ilmenite. This means that synthetic rutile from ilmenite plays an important role in the titanium industry. Less than 10 percent of the titanium concentrate is used in titanium metal production. The rest is used as titanium dioxide in pigments to increase opaqueness or intensity in paints, paper, and medicine.[4]

Sponge

Titanium sponge is the first commercial form of titanium metal that is refined from titanium ores. It is called "sponge" because of its porous, sponge-like appearance. Sponge is produced in various grades, with

---

[2] We discuss the cost structure of titanium production in detail later in this chapter.

[3] Ilmenite ores are used in iron production. They leave a TiO2-rich slag, which is usually upgraded to be used in titanium production.

[4] According to the *USGS Minerals Yearbook 2005*, worldwide production of titanium dioxide for the chemical industry is estimated at around 2.5 million tons per year and is continuing to grow.

varying levels of impurities. Higher-grade sponge is used in engine parts and man-rated static airframe parts; lower-quality sponge is used in commercial products, such as golf clubs.

Ingot

Titanium ingot is produced from titanium sponge, titanium scrap, or a combination of both. Titanium ingot is often an alloy, containing such metals as vanadium, aluminum, molybdenum, tin, and zirconium. Titanium alloyed with 6 percent aluminum and 4 percent vanadium, called Ti-6Al-4V, is most commonly used in the aerospace industry. Titanium ingot is produced in either a cylinder or a rectangular slab that may weigh several metric tons. It may be used for titanium castings or to produce mill products.

Mill Products

Mill products are produced from titanium ingot through such primary fabrication processes as rolling and forging. They are in the shape of billet, bar, plate, sheet, tube, and wire. These basic forms are the inputs to secondary fabrication. In secondary fabrication, titanium mill products are turned into finished shapes and components.

Production Processes

Titanium production requires complicated processes that are capital- and energy-intensive.[5] Refining the ore to titanium metal requires multistep, high-temperature batch processes. At the temperatures required for its reduction, titanium cannot be exposed to the atmosphere because its great affinity for oxygen, nitrogen, carbon, and hydrogen will make the metal brittle (Masson, 1955; Kraft, 2004). Therefore, either vacuum or inert gas metallurgy techniques are necessary to

---

[5] The details of the titanium metal production process described in this subsection are drawn from Hurless and Froes, 2002; DoD, 2004; Gerdemann, 2001; Kraft, 2004; TIMET 2005, 2006; and *USGS Minerals Yearbook 2005*.

reduce and process the metal. In addition, the hardness of the metal makes the machining process more difficult and time consuming.

Extracting Titanium Metal from Ore

Extracting titanium metal from ore requires multiple laborious steps. Titanium ores are chlorinated to produce titanium tetrachloride and then reduced with magnesium (called the Kroll process) or sodium (called the Hunter process) to form commercially pure sponge.[6] The Kroll process, which is the most common and least expensive process for producing titanium sponge, has four major steps. First, rutile concentrate or synthetic rutile (titanium slag) is chlorinated to form titanium tetrachloride and then distilled to remove metallic impurities such as iron, chromium, nickel, magnesium, and manganese. Second, the titanium tetrachloride is reduced with magnesium.[7] Third, the remaining magnesium and magnesium chloride are removed, most commonly by vacuum distillation. In this technique, heat is applied to the sponge mass while a vacuum is maintained in the chamber, causing the residue to boil off from the sponge mass. At the end of the process, the residual magnesium chloride is separated and recycled. Fourth, the sponge mass is mechanically pushed out of the distillation vessel, sheared, and crushed.

Producing Ingot from Sponge

Titanium sponge, titanium scrap, or a combination of both is melted together in an electric arc furnace to produce titanium ingot. On average, 40–50 percent of the raw material is titanium scrap.[8] In the aerospace industry, sponge is typically melted two or three times to produce an ingot. Titanium ingot may be used to produce mill products or

---

[6] See Gambogi, 2004; Kraft, 2004; and Gerdemann, 2001, for details of the titanium refining technology.

[7] The Hunter process is similar to the Kroll process except that it uses sodium instead of magnesium to reduce titanium tetrachloride ($TiCl4$), which is commonly referred to as "tickle."

[8] The percentage of scrap in feedstock for ingot varies over different final usage of the ingot. For example, no scrap is currently used for producing titanium ingot for the F-22.

titanium castings. Figure 2.1 shows the conversion of sponge to ingot using the vacuum arc remelting (VAR) process.

Titanium ingot is usually produced by either VAR or cold hearth melting. In VAR, the inputs undergo a first melt. The surface of the resulting ingot is ground to remove defects and contamination and the cleaned ingot is inverted and welded to a stub. The ingot is then melted again to improve homogeneity and dissolution of the alloying elements. Titanium ingot intended for high-stress and high-fatigue applications, such as engine rotors, is usually melted a third time.

Cost drivers for the melting process include the labor-intensive electrode preparation, the need for multiple melts, and the yield loss produced by intermediate and final conditioning.

Figure 2.1
Vacuum Arc Remelting Process for Converting Titanium Sponge into Ingot

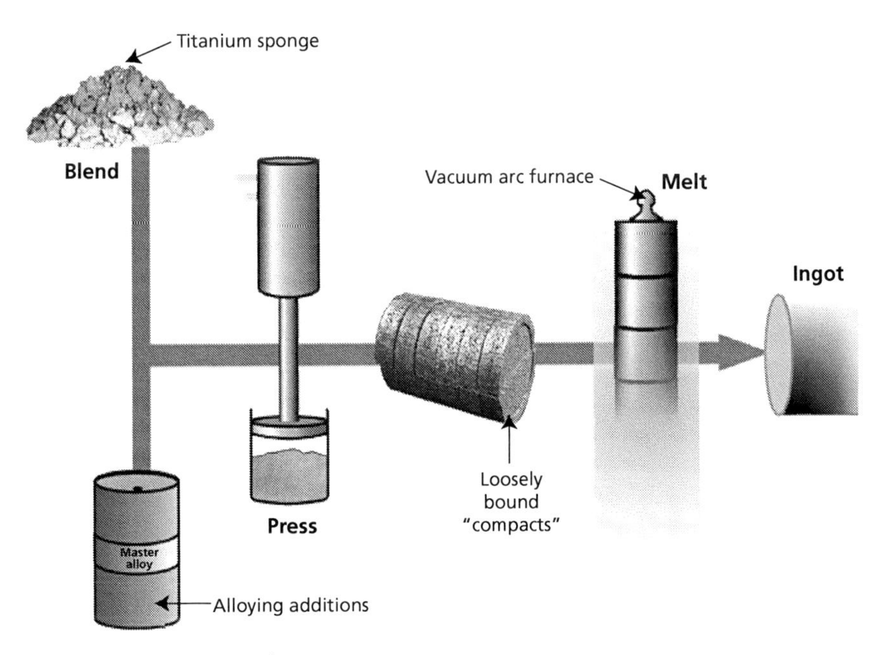

SOURCE: TIMET Corporation.
NOTES: The figure describes the process used at TIMET's Henderson, Nevada, plant. The process illustrated is not universal for all sponge production in the world.
RAND *MG789-2.1*

Cold hearth melting uses a water-cooled copper hearth to contain a "skull" of solidified titanium, which in turn holds a pool of molten titanium.[9] If gas plasma is used as the heat source, the process is called plasma arc melting (PAM). If an electron beam is used as the heat source, it is called electron beam melting (EBM). Cold hearth melting can substitute for VAR, but it also may be followed by a VAR melt to produce ingot for high-purity applications, such as aircraft engine rotors.

Cold hearth melting is a more cost-effective process than VAR because it includes fewer steps, can use more scrap, and allows a wider variety of scrap. Cold hearth melting can also cast rectangular slabs. Titanium plate for airframes can be produced more cheaply from rectangular slabs than from the round ingots created by VAR. However, cold hearth melting has some disadvantages compared with VAR—such as large surface areas for evaporation of volatile elements, the need for complex equipment, and batch processing by-products. For some higher-end products, such as aircraft engines, cold hearth melting cannot be used alone but should be combined with VAR.

Cold hearth melting will not be able to replace VAR completely in the near future. According to the *USGS Minerals Yearbook 2005,* about 20 percent of the U.S. titanium ingot capacity was produced by cold hearth melting that year, and the remaining 80 percent was produced by VAR.

Figure 2.2 displays the breakdown of the raw materials used to produce titanium ingot and mill products by TIMET, a U.S. titanium producer. The exact mix of titanium sponge, scrap, and alloy depends on the kinds of products to be produced and the quality of scrap available. TIMET both produced sponge in 2006 internally and purchased it on the market. The purchased quantity made up more than half of the total sponge it consumed. For titanium scrap used in 2006, TIMET generated material internally during production as well as material purchased on the market. The purchased quantity made up about 25 to 30 percent of the total scrap consumed. The breakdown

[9] See Figure 6 in Kraft, 2004.

Figure 2.2
Composition of Materials Used to Produce Titanium Ingot and Mill Products

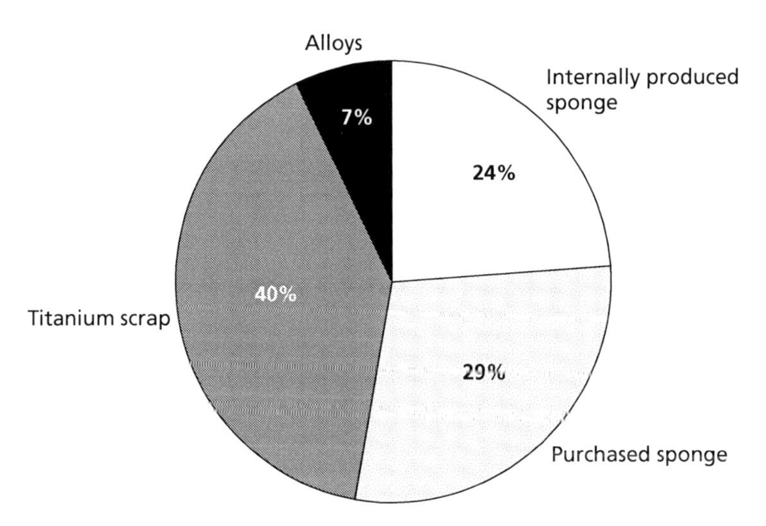

SOURCE: TIMET, 2006.
RAND *MG789-2.2*

between internally produced sponge or scrap and purchased sponge or scrap varies over different titanium ingot and mill product producers.

Primary Fabrication: Processing Ingot to Mill Products

Generic mill products such as plate, sheet, billet, and bar are produced from ingot or slab through a sequence of operations, such as forging and hot- or cold-rolling.[10] The fabrication process is complicated by the susceptibility of titanium metal to oxidation. Frequent surface removal

---

[10] *Sheet* is a flat-rolled product that is typically thinner than three-sixteenths of an inch and is produced by either hot- or cold-rolling. *Plate* is usually thicker than three-sixteenths of an inch and wider than 20 inches. It is produced by hot-rolling. *Billet* can have round, square, rectangular, hexagonal, or octagonal cross sections, with an area equal to or greater than 16 square inches and width less than five times its thickness. *Bar* has round, square, or rectangular cross sections with an area less than 16 square inches, thickness greater than three-sixteenths of an inch, and width less than or equal to 10 inches. *Forging* is a process in which metal is placed in a die and a compressive force is applied. Usually, the compressive force is in the form of blows from a power hammer or a press. *Rolling* is the most widely used method of shaping metals. It may be done while the metal is hot (hot-rolling) or cold (cold-rolling).

and trimming are required to eliminate surface defects. These operations are costly and involve significant yield loss.

Secondary Fabrication: Fabrication Parts from Mill Products

Production of finished titanium products from generic mill products requires a second set of steps, called secondary fabrication. Secondary fabrication includes a variety of processes, such as forging, extrusion, hot and cold forming, machining, and casting.[11] Selection of the fabrication process depends on the properties and shape of the final product. Secondary fabrication for titanium is similar to that for other metals, except for two main challenges. Titanium's hardness and reactivity slow the machining process and quickly wear down tools.[12] Also, machining generates large amounts of scrap, and the high price of titanium makes this waste a significant expense. Figure 2.3 illustrates the process of fabricating an aircraft part from a titanium ingot.

Scrap

Titanium scrap is a by-product of the fabrication processes used to produce titanium products. Titanium scrap is used as feedstock not only in titanium ingot production but also in ferrotitanium production. Because titanium scrap is a recycled good—not a manufactured

---

The process consists of passing the metal between pairs of revolving rollers. See Kraft, 2004; ITA, 2005a; and the Encyclopedia Britannica.

[11] *Extrusion* is a process by which long straight metal parts of one cross section are produced. The parts can be solid or hollow and can be round, rectangular, or other shapes. In *cold forming,* metal is placed under pressure, without adding heat, until a desired shape is achieved. Cold forming uses dies and punches to convert a specific metal of a given volume into a finished shaped part of the exact same volume. *Hot forming* is used in manufacturing industrial fasteners, such as bolts, screws, and rivets. In this process, heat is used to soften the metal, which is usually in the form of a sheet, bar, tube, or wire. Pressure is then used to alter the shape of the metal. *Machining* is a group of processes in which material is removed from a workpiece in the form of chips. It involves operations such as cutting, drilling holes, and grinding. In *casting,* liquid metal is poured into a mold, which contains a hollow cavity of the desired shape, and is allowed to solidify. The solid is then removed from the mold to complete the process.

[12] Here, *reactivity* concerns chemical reaction. Reactivity is the relative capacity of an atom, molecule, or compound to undergo a chemical reaction with another atom, molecule, or compound.

Figure 2.3
Converting a Titanium Ingot into an Aircraft Part

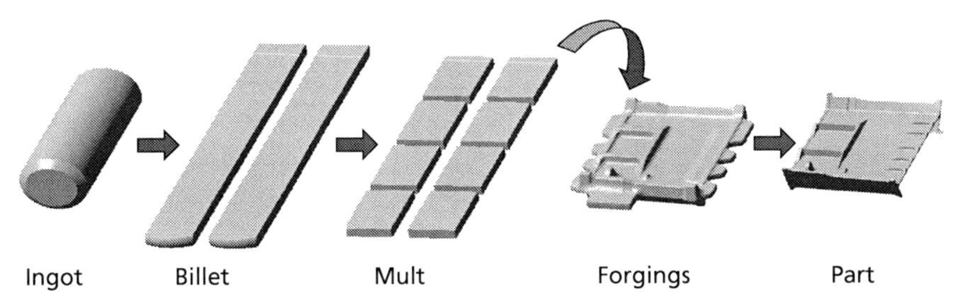

SOURCE: Wyman-Gordon.
RAND *MG789-2.3*

good—the supply of scrap depends on the consumption cycles of titanium end users. Not all scrap is reusable. For instance, machine turnings of different alloys that are not carefully segregated from other metals are too difficult to reuse. Once different metals and alloys are well segregated, typical contaminants such as oil and lubricants should be removed. Therefore, machine turnings are usually washed with soap and dried before being melted. Other scrap includes bulk solids.[13] Bulk solids can be reprocessed into ingot through the VAR process. Cold hearth melting can be used for machine turnings and smaller pieces of scrap, while VAR is less effective for small pieces. The ratio of scrap to sponge is determined by the customer's specifications for the ingot.

Ferrotitanium

Titanium also can be combined with iron to create ferrotitanium. Ferrotitanium is an alloy that is used in stainless steel and other specialty steel production for deoxidation, achieving a finer-grained structure, and controlling carbon and nitrogen. The two standard grades of ferrotitanium contain 40 percent and 70 percent titanium. Significant

---

[13] *Bulk solids* are large pieces of titanium created during the machining or the forging of titanium parts.

amounts of ferrotitanium, titanium scrap, and titanium sponge are consumed by the steel industry and other industries that produce alloys.[14]

Production Cost Structure

Titanium is expensive to refine, process, and fabricate. As shown in Table 2.1, titanium is much more expensive than aluminum and steel in all stages of production, including metal refining, ingot forming, and sheet forming (Hurless and Froes, 2002). In terms of the processing cost of materials per cubic inch, titanium is five times more expensive to refine than aluminum, and more than ten times more expensive than aluminum to form ingots and fabricate finished products.

Of all the stages of production, fabrication is the most costly, followed by extracting sponge from ore. Assuming that the Kroll process is used for sponge production and VAR is used for ingot production, Kraft (2004) calculated the cost composition of the conventional mill processing of a one-inch-thick titanium alloy plate. Fabrication is the largest cost factor, accounting for 47 percent of the mill product cost. Sponge production is the second largest cost, accounting for about one-third of the total. The ingot melting process makes up approximately 15 percent of the production cost, while rutile ore accounts for 4 percent.

Table 2.1
Cost Comparison of the Stages of Metal Production

Production Stage	Steel	Aluminum	Titanium
Metal refining	0.4	1.0	5.0
Ingot forming	0.6	1.0	10.7
Sheet forming	0.4	1.0	18.0

SOURCE: Hurless and Froes, 2002.

NOTE: Process costs were estimated in dollars per cubic inch of the relevant material and then normalized to the cost of aluminum.

[14] According to the *USGS Minerals Yearbook*, about 11,000 tons of titanium products were consumed in the production of steel and other alloys in the United States in 2005.

Refining Cost

Refining the ore into titanium metal is a costly, multistep, high-temperature batch process that is energy and capital intensive. Because of the high reactivity of titanium, an extraction process similar to that for aluminum has not yet been developed.[15]

Refining titanium metal into ingot is a slow, energy-intensive process that requires significant capital. VAR requires about 25 hours to melt an ingot, and the process must be done two or three times. The ingots are produced in three- to seven-ton batches, requiring a large space and involving many steps between each melt. Not only are the space and production equipment expensive, but the labor involved in moving batches adds significant cost (Gerdemann, 2001).

Fabrication Cost

The two main factors in the fabrication cost are the slow machining process and the high BTF ratio in part production.

The hardness that makes titanium so desirable also makes it more difficult to machine than traditional aluminum. This presents a challenge akin to that of machining high-strength steel. However, the process is complicated by titanium's high reactivity and low thermal conductivity. It is highly reactive and tends to wear tools very quickly, especially at higher temperatures. The low thermal conductivity means that high temperatures can be generated easily in the course of machining. Consequently, titanium must be machined at lower tool speeds, slowing production.

Titanium parts have very high BTF ratios, with most parts machined from large, solid pieces of metal. Because the raw material is so expensive, scrap represents a significant portion of the cost. As explained above, not all scrap can be reverted or recycled. In the fabrication process, a significant portion of the cost of a part is often left on the machining room floor.[16] Titanium producers and aircraft manufacturers try to recycle titanium scrap as efficiently as possible

---

[15] We discuss the development trends in titanium processing technologies in Chapter Six.

[16] It was not until recently that military aircraft programs started to pay much attention to titanium scrap recycling.

through various contractual arrangements and supply-chain management practices.

Buy-to-Fly Ratio

The BTF ratio depends on the form of the initial material (ingot, plate, sheet, etc.) and the fabrication process (forging, machining, casting, etc.). The BTF for titanium parts fabricated from plate or sheet is often more than 20 to 1—that is, to produce a part weighing only a pound may require more than 20 pounds of the raw metal (Younossi, Kennedy, and Graser, 2001). In machining, losses take the form of machine turnings, which can be reverted as scrap if properly handled and segregated from other materials.

A more common titanium manufacturing process is forging, which forms a nearly final-shaped (also called near-net-shaped) part from plate. In this process, ingot is rolled into plate and then subjected to extreme pressure and temperature, causing it to flow into a mold or die. The forging includes additional material called "cover," as well as tabs to easily attach it to a milling machine. Forgings also include an extension used as a test coupon to examine the metallurgic and mechanical properties of the material. The forging, flushing,[17] cover, tabs, and the test coupon have to be removed by machining. Most of the scrap resulting from this process is reverted.[18]

Cost-Saving Technical Changes

Titanium extraction technology has not progressed much since the Kroll process was commercialized in 1948. And the Hunter process is not any cheaper than the Kroll process. Several companies are developing improved processes that could produce titanium metal more cheaply and quickly than with the Kroll process. However, the com-

[17] *Flushing* is the extra amount of material that is extruded from the die during the forging process.

[18] Presentation to the authors by representatives of Wyman-Gordon Forgings, August 2007.

mercial viability of many of these processes is still unclear.[19] Whether there will be a breakthrough in titanium extraction technology in the near future, in the same way as the aluminum industry revolutionized its extraction process many years ago, is uncertain.

The technology for processing sponge into ingot has not changed much either. VAR has been the most commonly used process since its introduction in 1952. EBM and PAM are not used as widely as VAR. There are alternative approaches, such as investment casting and powder metallurgy, which aim to eliminate multistep melting and produce near-net-shape products. However, wide application of these new technologies to the aerospace sector may take a long time.

A major cost of technological change is certification of new materials. Standards are typically based either on judgments by a government body, such as the Federal Aviation Administration or the Air Force, or on standards set by the primary aircraft manufacturers. Within the Air Force, materials and processes must be certified separately for each program. The certification process typically lasts 18 to 24 months and requires extensive qualification processes. In the course of this process, a company must manufacture test articles and validate their properties at its own expense. The cost of this process prevents companies from attempting to certify materials until they are quite certain of their performance and properties. Consequently, a material must be used for several years in other applications before designers will consider it for aerospace uses.

Summary

The titanium industry produces a variety of products—titanium sponge, ingot, and mill products. They are mainly intermediate goods produced in the process of refining the titanium ore and melting and fabricating the metal. Titanium production requires compli-

[19] Commercial viability concerns whether commercialization of the technology would provide acceptable returns to investors in a reasonably foreseeable market and under reasonable operating conditions.

cated processes that are capital- and energy-intensive. Of all the stages of titanium production, fabrication is the costliest process, followed by extracting sponge from ore. Refining ore into titanium metal is a costly, multistep, high-temperature batch process. Because of the high reactivity of titanium, an extraction process similar to that for aluminum has not yet been developed. The two main drivers of titanium's high fabrication costs are its slow machining process and the high BTF ratio required for titanium part production. The hardness that makes titanium so desirable also makes it more difficult to machine than traditional aluminum. Titanium parts have very high BTF ratios, with most parts machined from large, solid pieces of metal. The BTF for a titanium part fabricated from plate or sheet is often more than 20 to 1—that is, a part weighing only a pound may require more than 20 pounds of the raw metal to produce. Chapter Six examines a number of emerging technologies that have the potential to address these various production cost drivers.

CHAPTER THREE

The Titanium Industrial Base and Other Market Characteristics

This chapter discusses the global sources of titanium sponge, the role of the supplier base, and other market characteristics—major buyers, substitutes and complements, and market price systems.

Geographic Distribution

About half of the world titanium sponge capacity in 2005 was located in Russia, Kazakhstan, and Ukraine, as shown in Figure 3.1. However, Japan had the largest titanium sponge production capacity, accounting for 33 percent of world capacity, followed by Russia at 25 percent, and Kazakhstan at 19 percent (International Titanium Association, 2005b). The United States had approximately 8 percent of global titanium sponge production capacity in 2005. China's sponge capacity is similar to that of the United States. However, China mainly produces titanium sponge for industrial uses, as its quality does not meet aerospace requirements. In 2004 and 2005, the Chinese steel industry consumed considerable amounts of titanium scrap and sponge, which drove up market prices for both commodities considerably and provided an incentive for construction of multiple new sponge plants in China.[1] In 2007, the combination of slightly reduced steel demand

[1] In 2004, China accounted for over 30 percent of world steel consumption.

Figure 3.1
Geographic Distribution of World Titanium Sponge Production Capacity, 2005

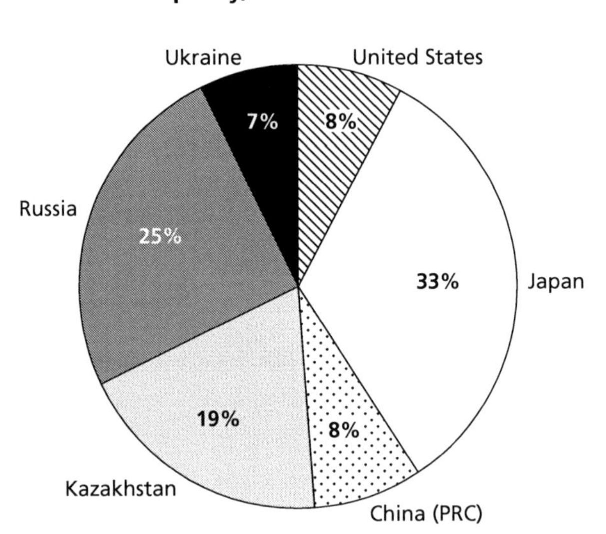

SOURCE: ITA, 2005b.
RAND *MG789-3.1*

and the significant increase in sponge production in China has had the reverse effect on the market.[2]

The United States is a net importer of titanium sponge. In 2005, approximately 60 percent of U.S. sponge consumption was imported. U.S. sponge imports originate mainly from Kazakhstan and Japan. As shown in Figure 3.2, Kazakhstan accounted for 53 percent and Japan accounted for 39 percent of U.S. sponge imports in 2005.

Major Producers

Like the steel and aluminum industries, the titanium market is an oligopoly. There are a limited number of titanium manufacturers in the world and market shares are concentrated in a small number of large producers.

---

[2] Authors' discussion with TIMET in August 2007.

Figure 3.2
U.S. Titanium Sponge Imports by Origin, 2005

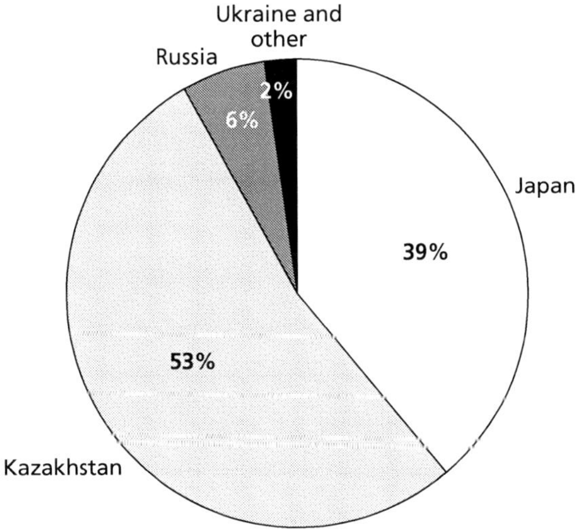

Ukraine and other
Russia
2%
6%
Japan
39%
53%
Kazakhstan

SOURCE: ITA, 2005b.
RAND *MG789-3.2*

One Russian company and three U.S. companies are the major producers of high-quality titanium metals for aerospace. The Russian company, VSMPO-Avisma (Verkhnaya Salda Metallurgical Production Association) is the world's largest titanium metal producer, with approximately 29 percent of the market share in global titanium shipments (Bush, 2006). The three major U.S. titanium metal producers are TIMET, Allvac (Allegheny Technologies Inc. [ATI]), and RMI Titanium Company (RTI International Metals, Inc.).[3] In the titanium market for industrial and emerging users, UNITI (a joint venture

[3] TIMET operates three melting plants, located in Henderson, Nevada; Morgantown, Pennsylvania; and Vallejo, California. ATI also has three melting plants: Richland, Washington; Bakers, North Carolina; and Albany, Oregon. RTI melting plants are located in Niles, Ohio, and Canton, Ohio. TIMET and ATI are integrated titanium manufacturers that produce a full spectrum of titanium products from titanium sponge to ingot and mill products. In 2007, RTI also announced it would build a new titanium sponge factory to be in operation by 2010.

between ATI and VSMPO), TIMET, RTI, and Japanese companies are the major producers.

Of the major U.S. producers, TIMET had approximately an 18 percent market share of the world's titanium industry shipments and an 8 percent market share of the world titanium sponge production in 2005.[4] TIMET is the only aerospace-engine-grade sponge producer in the United States and one of the largest sponge producers in the world.[5] The other major sponge producers are located in Russia, Kazakhstan, China, and Japan.

In addition to the three major U.S. producers mentioned above, there is one more titanium sponge producer and another titanium ingot producer in the United States (*USGS Minerals Yearbook 2005*). Alta Group (Honeywell International, Inc.) has a small electronics-grade high-purity sponge plant in Salt Lake City, Utah, which uses the Hunter process. Howmet Corporation (Alcoa, Inc.) also has some titanium ingot capacity in Whitehall, Michigan. Unlike the three major U.S. producers (TIMET, RTI, and ATI), these two smaller companies are not integrated producers.

Major Buyers

Major titanium buyers include the commercial and military aircraft manufacturing industry, the industrial equipment sector, and the consumer goods sector.

In the aerospace industry, titanium is mainly used in airframe components and jet engines. In an airframe, titanium is used in bulkheads, the tail section, landing gear, wing supports, and fasteners. In jet engines, titanium is used in blades, discs, rings, and engine cases. The

---

[4] TIMET, 2005.

[5] As of 2005, TIMET's sponge plant in Henderson, Nevada, could produce up to 8,600 tons per year through the Kroll process combined with vacuum distillation. In 2008, ATI started to produce aerospace-grade sponge, but not aerospace-engine-grade sponge.

commercial aircraft manufacturing industry is the largest single consuming market for titanium in both the world and U.S. markets.[6]

Titanium is used for industrial applications primarily due to its excellent corrosion resistance, which allows plants to reduce the maintenance cost and life-cycle costs of equipment. For industrial applications, "commercially pure" grade titanium is generally used.[7]

Industrial-sector buyers include a diverse set of producers of industrial equipment. The chemical and petrochemical industry is a leading buyer, using titanium for corrosion resistance in such equipment as heat exchangers, tanks, process vessels, and valves.[8] Other major industrial buyers include power plants, desalination plants, pollution control equipment producers, and the aluminum and steel industries.

Military buyers of titanium are mainly fighter programs in the United States and Europe, but the use of titanium in ground combat vehicle and naval applications is also growing. The relative importance of the military sector has been increasing recently due to the focus on light armaments and mobility, and this trend is expected to continue for at least the next several years.

Relatively small but high-growth-potential users of titanium are off-shore oil and gas production facilities, particularly for deep-water oil and gas fields; passenger cars, trucks, and heavy vehicles; geothermal facilities; architecture; medical devices and instruments; and golf clubs. Titanium producers often call these sectors emerging markets.

In 2006, the non-aerospace industrial equipment sector, consisting of various manufacturing industries, accounted for 50 percent of the titanium mill product demand in the global titanium market. However, the commercial aerospace industry alone accounts for 38

---

[6] Ti-6Al-4V is the titanium alloy most commonly used in aircraft manufacturing .

[7] There are two types of titanium, commercially pure titanium and titanium alloy. Titanium alloys contain other metals, such as aluminum and vanadium, and are used for jet aircraft engines, airframes, and other components. Commercially pure (CP) titanium is unalloyed titanium used in the power generation and chemical processing industries. CP grades 1 and 2 are most commonly used for industrial applications.

[8] A *heat exchanger* is a device that transfers heat from a hot to a cold fluid. Heat exchangers are widely used in fossil-fuel and nuclear power plants, gas turbines, heating and air conditioning, refrigeration, and the chemical industry.

percent of titanium mill product demand, followed by the military sector and emerging markets at 6 percent each (see Table 3.1).[9] Global consumption of titanium was estimated at 68,500 tons in 2005 and 75,000 tons in 2006.

In the United States, the aerospace industry accounted for 60–75 percent of the titanium sponge consumption over the past decade (see Figure 3.3) (USGS, *Historical Statistics,* 2008). The aerospace industry's share decreased from 75 percent in 1991 to 60 percent in 2004, but it rebounded to 75 percent in 2005 due to the dramatic increase in aircraft orders. In 2006, approximately 72 percent of titanium sponge was used in aerospace applications, and the remaining 28 percent was consumed in military armor and other industrial sectors (USGS, Mineral Industry Survey, 2006). Compared with the composition of global tita-

Figure 3.3
Aerospace Industry's Share of Titanium Sponge Consumption in the United States, 1975–2005

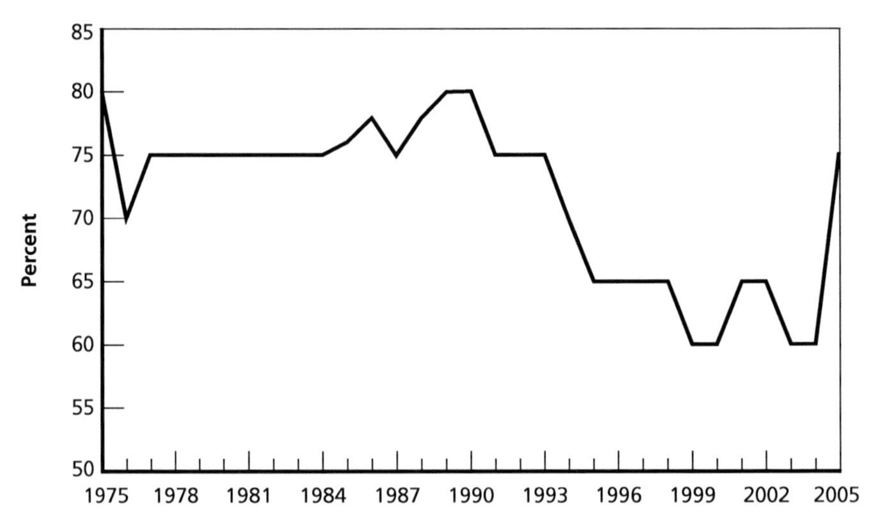

SOURCE: USGS, *Historical Statistics for Mineral and Material Commodities in the United States,* 2008.
RAND *MG789-3.3*

[9] The demand composition is calculated from world titanium mill product shipment statistics in TIMET, 2005.

nium demand, aerospace buyers constitute a considerably larger share of the U.S. market.[10] In the global market, aerospace demand accounts for only 44 percent of the global titanium demand, as shown in Table 3.1.

The dominance of aerospace buyers in the U.S. titanium market can also be observed in the revenues of TIMET. In 2006, 57 percent of TIMET's revenues came from commercial aerospace buyers, 15 percent from the military, 17 percent from industrial equipment producers and emerging markets, and 11 percent from other industries (Figure 3.4).[11]

Table 3.1
World Titanium Mill Product Shipments by End-User Sector, 2005–2006

	2005		2006	
Sector	Quantity (tons)	Share (%)	Quantity (tons)	Share (%)
Commercial aerospace	24,000	35	28,500	38
Military aerospace	5,200	8	4,400	6
Industrial equipment	35,600	52	37,600	50
Emerging markets	3,700	5	4,500	6
Total mill product shipments	68,500	100	75,000	100

SOURCES: TIMET, 2006; other historical industry data from titanium melters, forgers, and casters.

NOTE: Non-aerospace military users are included in the emerging markets.

[10] The aerospace share of sponge consumption does not necessarily reflect its share of titanium mill product demand, because the raw material mix is different between titanium products for the aerospace sector and those for the industrial sector. We use the share of sponge consumption only as a proxy, since data for the aerospace share of titanium mill product consumption in the United States are not available.

[11] TIMET's revenue structure may not represent the general revenue structure of titanium producers in the United States. We display the TIMET case here because comparable data from other major U.S. producers, such as ATI and RTI, are not available.

Figure 3.4
Sectors to Which TIMET's Titanium Mill Products Were Shipped, 2006

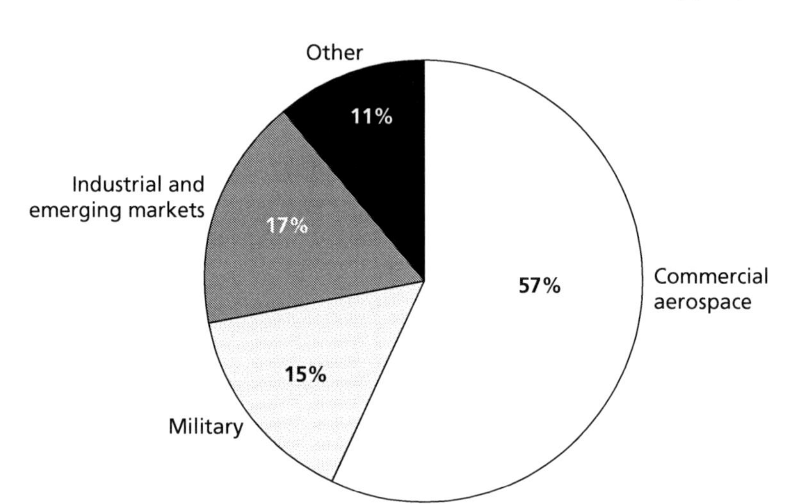

SOURCE: TIMET, 2006.
NOTE: "Other" revenue includes sales of titanium fabrication, titanium scrap, and titanium tetrachloride.
RAND *MG789-3.4*

Substitutes and Complements

Other materials can be used as substitutes for titanium in various applications (USGS, Mineral Commodities Summary, 2007). In high-strength applications, titanium substitutes include composites, aluminum, steel, and high-performance alloys. In corrosion-resistance applications, aluminum, nickel, stainless steel, and zirconium alloys may substitute for titanium. However, these are not perfect substitutes. Even if the relative prices of the substitutes change, the degree of substitution is usually limited by the requirements of the specific application. Especially in military aircraft applications, performance considerations often overrule cost concerns.

Titanium and composites complement each other in airframe applications. Among aerospace metals, titanium is very compatible with carbon fiber reinforced polymer (CFRP) composites (Fanning, 2006). Titanium's stiffness and its coefficient of thermal expansion are similar to that of CFRP composites. Titanium is electrochemically compatible with carbon and resists galvanic corrosion, in contrast to aluminum, which tends to corrode when placed next to a CFRP composite.[12] These compatibilities make titanium an excellent substructure material for maximizing the effectiveness of composite parts. According to industry experts, usage of titanium in commercial and military airframes increased significantly over the last decade, partly due to the increasing use of CFRP composites.[13]

Market Price

Oligopolistic Price

Like other metal industries, such as the steel and aluminum industries, the titanium market is an oligopoly because there are a limited number of titanium manufacturers in the world. In an oligopolistic market, the small number of suppliers often results in prices that are higher than those in a competitive market.

Market Size and Market Risks

Unlike other metals markets, the titanium market is extremely small. World titanium sponge production capacity was only 110,000 tons in 2005, whereas world crude steel production for the same year was 1.1 billion tons.[14] Given the small market size and highly concentrated buyers and suppliers, the titanium market is more exposed to turbulence caused by supply and demand shocks than are large raw material markets with diversified buyers and suppliers.

[12] Galvanic corrosion results when two materials with dissimilar electrical potentials come into contact. Over time, ions from one material will migrate to the other, gradually creating corrosion.

[13] Authors' discussion with Boeing in July 2007.

[14] Titanium sponge capacity data are from USGS, Mineral Commodity Summaries, 2006. Crude steel production data are from the International Iron and Steel Institute (IISI).

Spot Market Versus Long-Term Contracts

Titanium prices are based on both spot market transactions and longer-term contracts. Spot market transactions are basically delivery-on-order agreements in which the purchaser pays the current price for titanium. Titanium is not traded on a fully developed auction market such as the London Metal Exchange.[15] Contracts for less than three years are considered short-term contracts; contracts for three years or more are considered long-term. Major buyers of titanium often prefer five- to ten-year contracts.

The coexistence of spot market prices and contract prices is a common characteristic of raw material markets. Due to the two different price systems, there are multiple prices for the same good in the same market at a given time. Spot market prices adjust quickly to supply-and-demand shocks, whereas long-term contract prices are more rigid.

Compared to other raw materials, such as aluminum, steel, copper, and oil, titanium spot market transactions are less prevalent, and the titanium spot market has only recently become significant. Although the exact size of this market is unknown,[16] industry experts estimate that it usually accounts for at least 10 percent of transactions and can easily account for one-third to more than one-half of all transactions.[17]

The limited volume of titanium spot market transactions—compared to other raw materials—is partly due to the limited size of the titanium market in general, the characteristics of downstream industries, and the nature of price shocks.[18]

---

[15] The London Metal Exchange is the world's largest market in options and futures contracts on metals. The metals exchanged in the London Metal Exchange include aluminum, copper, nickel, tin, zinc, and aluminum alloys. Refer to the London Metal Exchange Web site, 2008, for details.

[16] The size of the titanium spot market is unknown because titanium contracts are not traded through an organized metal exchange. There is no institute or agency that compiles titanium transaction data for public use.

[17] The percentage of spot market transactions in titanium varies depending on business cycles and when long-term contracts expire, according to industry experts.

[18] *Downstream industries* are industries that buy titanium.

Conversely, long-term contracts still dominate the titanium market, accounting for approximately 50 percent or more of the transactions by major titanium producers. The aerospace industry is the dominant downstream industry in which the major buyers are fixed and production lead time is significant. Three U.S. companies and one Russian company are the major producers of high-quality titanium metals for aircraft manufacturing.[19]

Sources of underlying price fluctuations may provide another explanation for the prevalence of long-term contracts in the titanium market. According to Hubbard and Weiner (1985), when demand shocks are more significant relative to supply shocks, it is more likely there will be greater contracting and price rigidity.

Long-term contracts usually stipulate conditions such as minimum annual quantity, minimum share of the customers' titanium requirements, prices determined by an agreed-upon formula, and price adjustments for raw material and energy cost fluctuations. Long-term contracts reduce price volatility for the buyer while securing a base level of revenue for the supplier throughout the business cycle. In 2005, approximately 49 percent of TIMET's sales revenue was from customers under long-term contracts, such as Boeing, Rolls-Royce, and United Technologies (TIMET, 2005).

Import Tariffs

As of 2007, imports of the titanium mill product called wrought titanium are subject to a 15 percent tariff if the country supplying the import has normal trade relations with the United States (USGS, Mineral Commodities Summary, 2007). If wrought titanium imports are from countries that do not receive normal trade relations treatment, the tariff is 45 percent. The tariff on titanium sponge and ingot, or unwrought titanium, is 15 percent regardless of trade relations treatment.

The tariff on wrought titanium may unilaterally protect domestic mill product producers against foreign producers. However, the tariff on unwrought titanium would have asymmetric effects on domestic pro-

[19] Details of major suppliers will be discussed later.

ducers. Domestic titanium ingot producers that use imported sponge pay 15 percent more for raw materials than their domestic competitors who do not rely on imported sponge. As mentioned above, approximately 60 percent of U.S. titanium sponge consumption depended on imports in 2005.

These tariffs influence the titanium mill product market price by increasing costs for domestic mill product producers who depend on imported raw materials. As a result, domestic producers who do not depend on imported raw materials have higher profit margins, assuming all other factors remain equal. However, if their final products are exported, the import tariff paid can be refunded through duty drawback.

Summary

The United States had approximately 8 percent of the global titanium sponge production capacity in 2005. That year, the United States imported approximately 60 percent of the sponge it consumed—and 92 percent of U.S. sponge imports came from Kazakhstan and Japan. There are a limited number of titanium manufacturers in the world and market shares are concentrated in a small number of large producers. One Russian company and three U.S. companies are the major producers of high-quality titanium metals for aerospace. Major titanium buyers include the commercial and military aircraft manufacturing industry, the industrial equipment sector, and the consumer goods sector. In the United States, the aerospace industry has accounted for 60 to 75 percent of the titanium sponge consumption over the last decade.

The titanium market is very small—only one ten-thousandth the size of the steel market. Given its small size and highly concentrated buyers and suppliers, the titanium market is more exposed to turbulences caused by supply and demand shocks than are large raw material markets with diversified buyers and suppliers. Compared with other markets, such as aluminum, steel, copper, and oil, titanium spot market transactions are less prevalent. However, industry experts estimate that the size of the titanium spot market varies from 10 percent to

50 percent, depending on business cycles and contract turnover in the industry. In addition, supply and demand shocks will become evident first in the spot market.

CHAPTER FOUR

Supply-Side Drivers of Titanium Price Fluctuations

On the supply side of the market, what factors drive titanium price fluctuations? In this chapter, we examine raw material availability and price trends, the responsiveness of capacity expansion to demand, production cost drivers and trends, and other supply-side drivers, such as government regulations.

Availability and Price Trends of Raw Material

Raw material, such as sponge and scrap, account for approximately 40 percent of the cost of titanium mill product production. Therefore, fluctuations in titanium mill product prices are often triggered by supply shocks in raw materials.[1] In fact, prices of titanium sponge and scrap started increasing sharply even before the commercial aircraft order surge in 2005 and 2006. As shown in Figure 4.1, average import

[1] In addition to the impact of sponge and scrap availability, one may ask whether the titanium dioxide pigment market has influenced the volatility of titanium metal prices. We found that the titanium dioxide pigment price and consumption have been quite stable in recent years. Annual average growth in consumption between 2003 and 2007 was less than 1 percent; the PPI for titanium dioxide pigment increased 3.3 percent annually in the same period.

Figure 4.1
Annual Inflation Rates of Titanium Sponge and Scrap Prices, 1994–2004

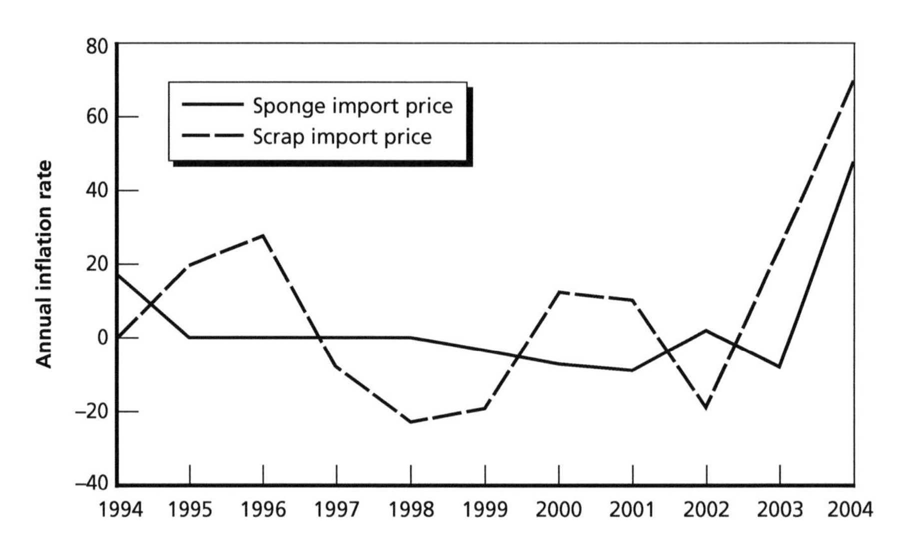

SOURCE: USGS, *Historical Statistics for Mineral and Material Commodities in the United States*, 2008.
RAND *MG789-4.1*

prices of titanium scrap and sponge increased 71 percent and 49 percent, respectively, in 2004.[2]

Sponge and Scrap Shortage

The heightened demand for ferrotitanium, titanium scrap, and titanium sponge from carbon and stainless steel production brought an added raw material shortage in 2003–2005. This cross-market substitution effect was significant, given the fact that the steel market is 10,000 times as large as the titanium market.

Sponge-capacity expansion requires a long lead time, so the gap between supply and demand needed to be filled with titanium scrap in the short run. However, titanium scrap was in extreme short supply,

---

[2] The import price of titanium sponge increased from $2.72–$3.95 per pound in 2003 to $3.55–$6.44 per pound in 2004. The scrap import price increased from $1.61 per pound in 2003 to $3.80–$4.00 per pound in 2004. Import prices of sponge and scrap were obtained from USGS, *Historical Statistics for Mineral and Material Commodities in the United States*, 2008.

especially from 2003 to 2005. Since scrap is a recycled good, the supply of scrap depends on titanium metal consumption. Scrap supply was extremely limited from 2003 to 2005 because aircraft production bottomed out in 2003.

On the other hand, scrap demand surged as growing carbon steel and stainless production caused a surge in ferrotitanium demand. As mentioned above, ferrotitanium is used in the steel production for reduction, achieving a finer-grained structure and controlling carbon and nitrogen content. In 2003–2004, world steel prices increased dramatically, mainly driven by China's strong demand and world economic recovery. The ferrotitanium PPI increased 82 percent from 2003 to 2005.[3] The phenomenal increase in the price of a substitute for titanium scrap further exacerbated the titanium scrap shortage.

Titanium scrap prices peaked in 2006. The scrap shortage situation has improved since 2006, as aircraft production increased significantly (thus producing more titanium scrap) and the price of steel products stabilized after 2005.[4]

The scrap shortage and ferrotitanium price surge induced sharp increases in titanium sponge prices because sponge, scrap, and ferrotitanium substitute for each other in the production of steel and other alloy industries. However, the world titanium sponge capacity in 2004 was 22 percent lower than the peak capacity in 1997. Given that an expansion of the titanium sponge production capacity would necessitate building a complete factory and would require a high capital investment and a long lead time, it was not possible for titanium producers to quickly expand the sponge supply to meet the unexpected surge in demand.[5] In addition, titanium sponge had not been a lucrative business since the end of the cold war in the early 1990s. Since the titanium raw material supply was already tight in 2004, the additional

---

3 Ferrotitanium buyers are insensitive to price because ferrotitanium accounts for less than 1 percent of steel production cost.

4 After 2005, China became a net exporter of steel.

5 Ingot and mill product capacity can be expanded incrementally by adding more furnaces, opening up a mothballed furnace, or utilizing steel mill facilities. However, sponge capacity expansion needs a complete factory.

demand shock from record levels of commercial aircraft orders in 2005 and 2006 further amplified the raw material shortage.[6]

Titanium spot market prices were especially volatile. For example, the spot market price of the titanium ingot used in aerospace manufacturing quadrupled in two years, from 2003 to 2005.[7]

Depletion of U.S. Titanium Sponge Stockpile

Titanium is regarded as a strategic material by the U.S. government. Under the terms of the Strategic and Critical Materials Stock Piling Act,[8] the Defense National Stockpile Center (DNSC) of the Defense Logistics Agency (DLA) was directed to maintain a large stockpile of titanium sponge, especially during the cold war. The purpose of the stockpile was to minimize dependence on foreign sources of strategic and critical materials in times of national emergency. After the cold war, the size of the stockpile was reduced significantly but was still maintained at more than 33,000 tons (see Figure 4.2). The size of the stockpile could cover total domestic consumption of titanium sponge for at least one year, even during peak consumption.

However, in 1997, Congress authorized the disposal of the stockpile at the DNSC.[9] The stockpile declined rapidly after that and was finally depleted in 2005, as shown in Figure 4.2 (USGS, *Minerals Yearbooks*).

Russia used to maintain a national stockpile of titanium sponge, which was also depleted recently. Sales from the national stockpiles can no longer serve as a buffer against spikes in sponge demand. As a consequence, unexpected surges must be met by production and industry stocks.

---

[6] We discuss the dramatic increase in commercial aircraft orders in recent years and its impact on the titanium market in the next chapter.

[7] According to Metalprices.com, the spot market price of titanium 6Al-4V increased from $5 per pound in the first quarter of 2003 to $20 per pound in the third quarter of 2005.

[8] 50 U.S.C. Section 98 et seq.

[9] National Defense Authorization Act for Fiscal Year 1998 Section 3304, "Disposal of Titanium Sponge in National Defense Stockpile."

Figure 4.2
U.S. Titanium Sponge Inventory Stocks, 1990–2006

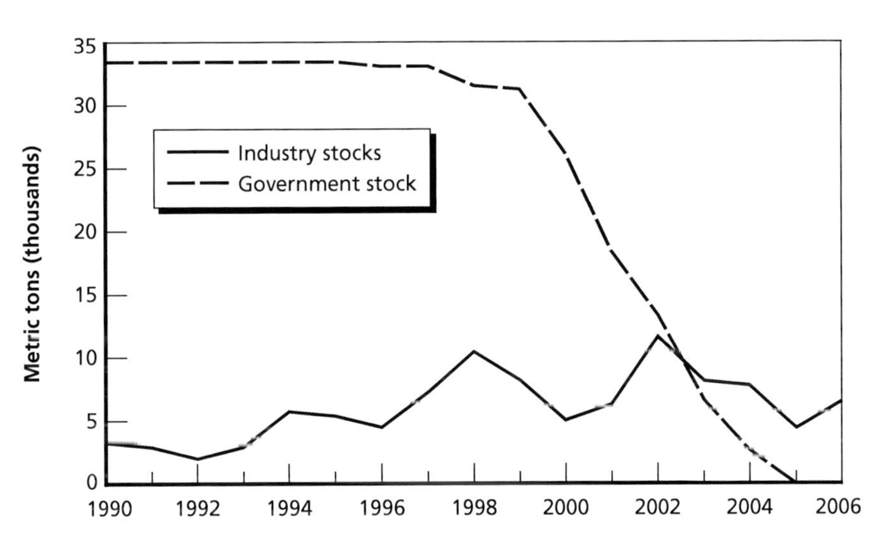

SOURCES: USGS, Mineral Industry Surveys and *Minerals Yearbook*, various years.
RAND *MG789-4.2*

The depletion of the titanium sponge stockpile maintained by the DLA coincided with the beginning of the shortage of sponge and scrap, accelerating the stockpile's decline. This was because major titanium producers had aggressively used the sponge stockpile as a substitute for scrap and ferrotitanium. The stockpile depletion significantly influenced the titanium market in the critical years of 2004–2006.

Responsiveness of Production Capacity to Demand

The response of the titanium metal supply lagged behind the unexpected surge in demand for several reasons. One source of sluggishness was the fabrication lead time necessary for mill products. Another was the 12–18 months needed to add more furnaces for ingot production.[10]

[10] According to industry experts, it takes about nine months to add the new furnace itself. In addition, securing permits to meet environmental, financial, and other regulations takes at least a few extra months, and often several months.

More important, sponge capacity could not be expanded without an additional factory, which would require an investment of $300 million to $400 million and take about three years to build. Right before the recent demand surge, titanium producers had suffered several years of bad business, and some of them had been at the verge of bankruptcy. As a result, titanium metal producers hesitated to invest in capacity expansion until they could be assured that increased demand would continue for at least the next several years.

Excess Production Capacity of Titanium Sponge Until 2004

After the cold war ended, the world had an excess of sponge production capacity. Sponge import unit values declined significantly from the early 1990s until the recent demand surge beginning in 2004 (Figure 4.3). In 1998 constant dollars, the sponge import price in 2003 was only half the 1991 price.

Adjusting to the market downturn, world titanium sponge production capacity decreased until 2001, as shown in Figure 4.4. World titanium sponge production capacity in 2004 was only 78 percent of that in 1997. However, the sponge capacity reduction in the United States was rather extreme. U.S. titanium sponge capacity in 2004 was only 40 percent of that in 1997. Until the recent surge in titanium demand, sponge production was not an attractive business to U.S. titanium producers.

At the beginning of the titanium sponge shortage in 2004, world titanium sponge producers first met increased demand by using more of their idle capacity rather than expanding it. As Figure 4.4 shows, there was no significant expansion in sponge capacity in 2003 and 2004. In 2005 and 2006, however, capacity, production, and capacity utilization rates all increased significantly in the rest of the world.[11] According to industry experts, world sponge capacity utilization was more than 90 percent in 2006, and production capacity for ingot and mill product was almost fully utilized in that year.

---

[11] The sponge production capacity utilization rate is calculated as production divided by capacity.

Figure 4.3
U.S. Titanium Sponge Import Price, 1985–2004

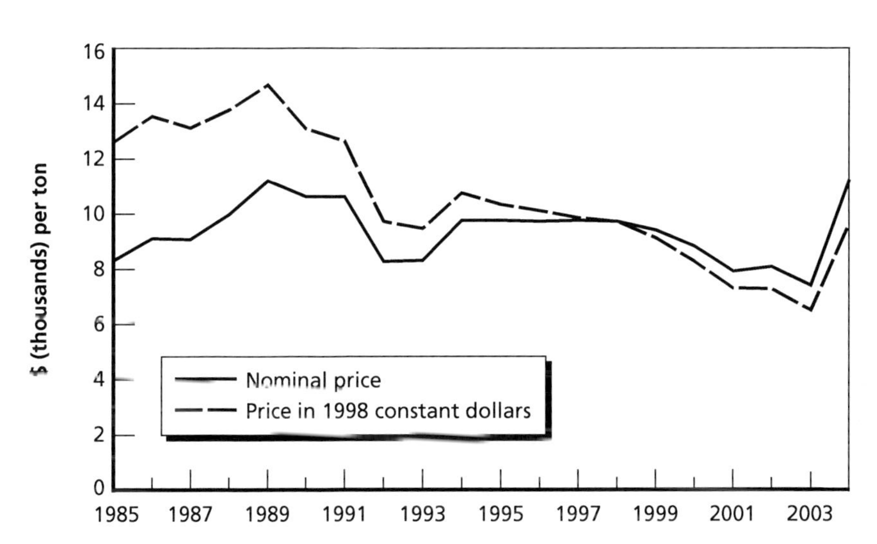

SOURCE: USGS, *Historical Statistics for Mineral and Material Commodities in the United States*, 2008.
RAND *MG789-4.3*

Titanium Sponge Production Capacity Expansion After 2004

Sponge producers expect that the increased demand for titanium is not a transient shock but a persistent phenomenon that will last for at least several years. Many believe the heightened demand for titanium is based on growth in the commercial aircraft industry, increasing military demand, and growing industrial demand from developing countries such as China and India.[12]

World titanium sponge capacity grew 10 percent in 2005 and 15 percent in 2006 (USGS, Mineral Commodities Summary, various years). The total world sponge capacity is estimated to have increased to more than 130,000 tons per year by the end of 2006. Sponge producers in Japan, China, and Russia are leading the recent expansion, accounting for 77 percent of the world sponge capacity increase in 2005 and

[12] See Holz, 2006, and other presentations at the International Titanium Association (ITA) Conference, San Diego, California, 2006.

Figure 4.4
World Titanium Sponge Production Capacity and Production Trends, 1995–2006

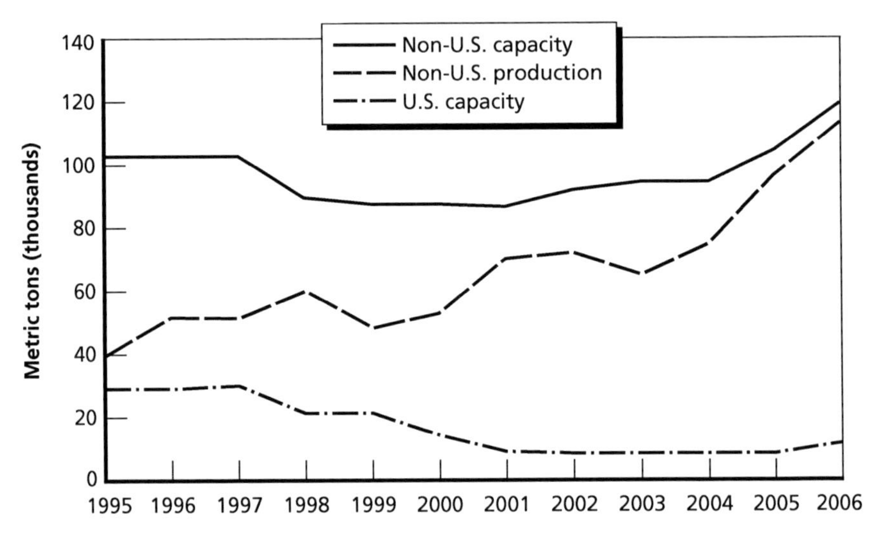

SOURCE: USGS, Mineral Commodities Summary, various years.
NOTES: Sponge production data for the United States are proprietary and not publicly available. Non-U.S. producers are China, Japan, Kazakhstan, Russia, and Ukraine.
RAND *MG789-4.4*

2006 (Figure 4.5). U.S. sponge capacity was estimated to have increased 35 percent in 2006, reaching 12,300 tons per year.

Other Supply-Side Factors

Entry and Exit

Before the recent price surge, U.S. titanium sponge production capacity decreased nearly 70 percent between 1995 and 2004. Many U.S. titanium metal producers exited the market or merged with large companies in the last decade. The industry's excess capacity after the cold

Figure 4.5
Titanium Sponge Production Capacity Trends by Country, 1995–2006

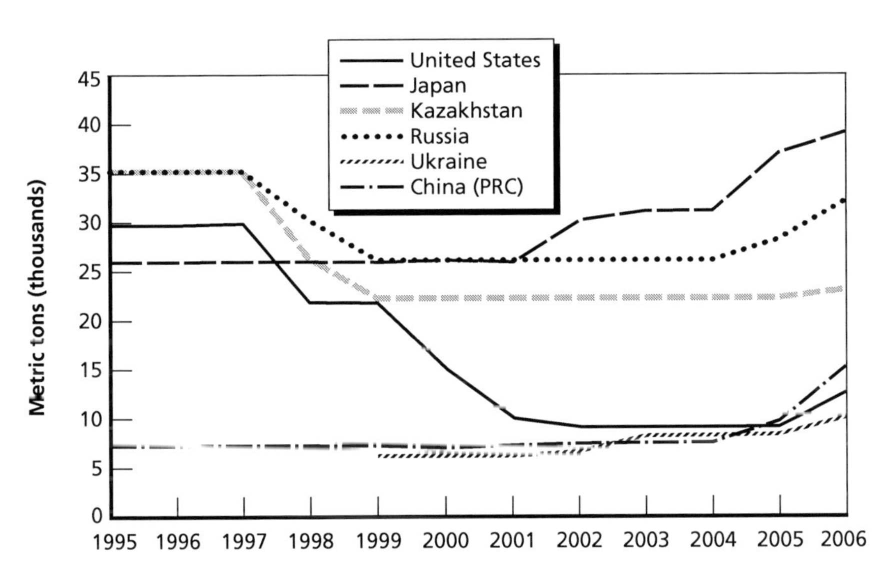

SOURCE: USGS, Mineral Commodities Summary, various years.
RAND *MG789-4.5*

war and its cyclical nature were among the major challenges to U.S. titanium producers.[13]

The number of titanium sponge and ingot producers in the United States decreased from 11 in 1995 to only five in 2005 (USGS, *Minerals Yearbook: Titanium,* various years). In 1994, there were two titanium sponge producers that also produced ingot, Oregon Metallurgical Corp. (Ormet) and TIMET; nine other ingot producers; and about 30 producers of titanium mill products and castings. In 2005, there were only five companies that produced either sponge or ingot: TIMET produced both sponge and ingot, Alta Group produced sponge, and Allvac, RMI Titanium, and Howmet Corp. produced ingot.

Since there was already excess capacity for more than a decade through 2004, no new integrated producers entered the U.S. titanium

[13] We discuss the cyclical nature of the industry in the next chapter.

metal manufacturing industry during that time.[14] Steel and aluminum metal producers that maintain forging, rolling, and finishing facilities could modify their plants to produce titanium mill products. However, industry experts do not believe that a completely new integrated titanium producer will enter the market in the near future, given the small size of the market, the cyclical nature of the business, and large capital commitment needed.

U.S. Titanium Metal Production Capacity Trends

U.S. titanium sponge production capacity declined 70 percent between 1995 and 2004, and U.S. sponge consumption fluctuated between 17,100 tons and 31,300 tons during the same period. This volatile demand was mainly accommodated by adjusting sponge imports (see Figure 4.6).

In contrast, U.S. ingot capacity increased 39 percent during the same period, from 61,400 tons in 1995 to 85,300 tons in 2005 (Figure 4.7). Most of this expansion occurred from 1996 to 1998. Between 1995 and 2005, ingot export prices peaked in 1999 at $18,500 per ton, and then declined to $12,700 per ton in 2003. From 1998 to 2003, ingot capacity utilization averaged about 61 percent of the peak utilization rate, which came in 1997.

The increase in ingot capacity was partly due to changes in production technology. As new melting technologies, such as EBM and PAM, were adopted, new types of furnaces were added to the existing capacity. Additionally, when ingot production peaked in 1997, U.S. titanium producers were optimistic that future demand would go even higher. As a result, they increased ingot production capacity 15 percent between 1997 and 1998. Instead, the market began a downturn in 1998 and the capacity utilization rate decreased 14 percent from 1997 to 1998. The industry waited until 2004 before the capacity utilization rate rebounded.

---

[14] In February 2008, American Titanium Works LLC, led by a Chicago attorney and former titanium industry executives, announced a $250 million investment to build a new production facility for titanium ingot and mill products.

Figure 4.6
U.S. Titanium Sponge Capacity, Consumption, and Imports, 1994–2005

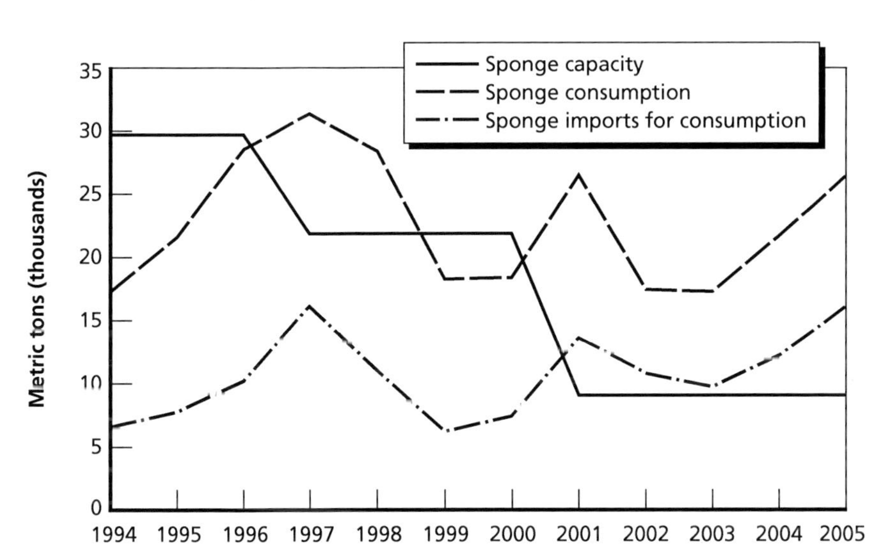

SOURCE: USGS, Mineral Industry Surveys, various years.
NOTE: Estimated operating capacity is based on seven-days-per-week full production.
RAND *MG789-4.6*

With the heightened demand for titanium in recent years, ingot production is operating at nearly full capacity, and many titanium producers added new furnaces in 2007. For example, TIMET's Morgantown, Pennsylvania, plant added two new EBM furnaces and one VAR furnace to its existing three EBM furnaces and one VAR furnace.

Berry Amendment

Titanium is one of the specialty metals defined by the Specialty Metals Clause in the Defense Federal Acquisition Regulation Supplement.[15] Specialty metals include titanium and titanium alloys, zirconium and zirconium base alloys, certain steel alloys, and other metal alloys. The Specialty Metals Clause implements the requirements of the Berry Amendment, which requires the Department of Defense (DoD) to

---

[15] Defense Federal Acquisition Regulation Supplement (DFARS) 22.225-7014, "Preference for Domestic Specialty Metals" clause. See also the FY 2007 Defense Authorization Act, 10 U.S.C.A. Section 2533b(i).

Figure 4.7
U.S. Titanium Ingot Capacity and Capacity Utilization Rate, 1994–2005

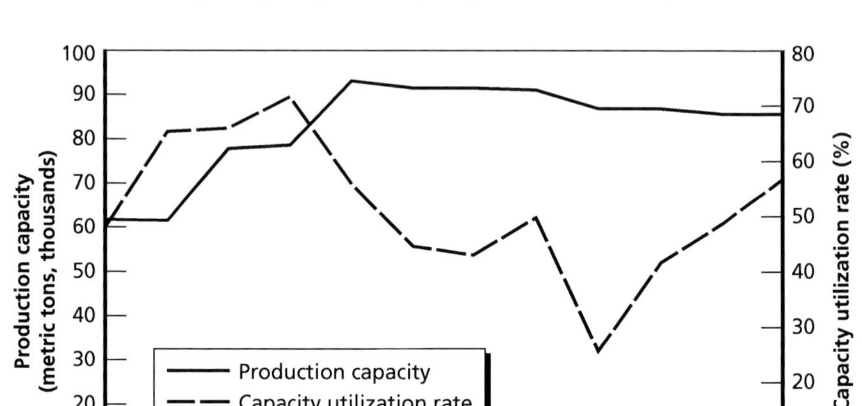

SOURCE: USGS, *Minerals Yearbook*, various years.
NOTES: Capacity utilization rate is calculated as production divided by capacity. Estimated capacity is based on seven-days-per-week full production. The ingot capacity of each furnace varies over product mix and process mix of the melting process.
RAND *MG789-4.7*

procure food, clothing, fabrics, and specialty metals from domestic sources.[16] Its origin goes back to the Fifth Supplemental DoD Appropriations Act of 1941.[17] The law was written to protect the domestic industrial base in case of a national emergency. According to the Berry Amendment, specialty metals incorporated in articles delivered under DoD contracts must be "melted or produced" in the United States.

In 2006, Congress modified the specialty metals restrictions of the Berry Amendment as part of the John Warner National Defense Authorization Act for Fiscal Year 2007.[18] To clarify the requirements, the act separated the Berry Amendment specialty metals restrictions from the

---

[16] Public Law No. 77-29, 55 Stat. 123 (1941).

[17] See Chierichella and Gallacher, 2004, for a history of the Berry Amendment.

[18] Public Law No. 109-364, Section 842, 120 Stat. 2083, 2335.

provisions applicable to other commodities.[19] Under the revised law, specialty metals should be melted or produced in the United States

- when specialty metals are "purchased directly" by the government or prime contractors
- when the procurement of end items and components of the six major systems—aircraft, missile and space systems, ships, tank and automotive items, weapon systems, and ammunition—contains specialty metals. For the six major systems, which include most military uses of titanium, the domestic source restrictions also apply to subcontractors. As a result, the Berry Amendment restrictions for titanium often apply to both prime contractors and subcontractors.

The Berry Amendment allows several exceptions to the specialty metals procurement restrictions, including the domestic nonavailability exception, the urgency exception, the qualifying country exception, and the de minimis exception, which applies to electronic components whose metal content does not exceed 10 percent of their value.[20] However, industry leaders still regard Berry Amendment requirements as quite restrictive. A coalition of industries called the Berry Amendment Reform Coalition insists that for the defense industry, the cost of compliance with the specialty metals provisions outweighs the benefit provided to the comparatively few domestic companies that produce specialty metals.[21] Chierichella and Gallacher (2004) even raise the possibility of reverse discrimination against domestic titanium producers, arguing that the qualifying-country exception of the Berry Amendment may "allow French and German companies to deliver Russian titanium to DoD" while domestic companies cannot. Since late 2007, Congress and the industry coalition have worked together

---

[19] The Berry Amendment specialty metal restrictions are in 10 U.S.C. Section 2553b.

[20] Refer to Churchill and Weinberg, 2007, for details regarding exceptions.

[21] The Berry Amendment Reform Coalition included many industry associations, such as the Aerospace Industries Association and the National Defense Industrial Associations. See Berry Amendment Reform Coalition, 2006.

to balance defense industrial base protection concerns against the cost of regulation.[22]

The effect of the Berry Amendment is still controversial. Supporters usually advocate a national security point of view by arguing the Berry Amendment protects a critical industrial base, making business viable in times of peace and war. Opponents often proffer an economic perspective, pointing out that the law can undermine free-market competition.

The Berry Amendment applies only to defense procurement, which is much smaller than the commercial market. However, if the surges in defense demand and commercial demand for titanium overlap, the Berry Amendment may further constrain the amount of available titanium in the United States and significantly increase production lead times. This may constitute a potential price driver—at least in the short run, given the limited number of domestic titanium producers, the sluggish nature of production capacity expansion, and the unlikely entry of new titanium producers in the near term.

China's Impact on Titanium Prices

Even though China's direct consumption of titanium is relatively small, it influenced titanium prices significantly during the recent market turmoil of 2003–2006. World steel consumption peaked in 2004, driven by world economic recovery and China's rapid growth in steel consumption, which caused a rapid increase in the prices of ferrotitanium at a time when there was an extreme shortage of titanium scrap (2003–2005). This cross-market substitution effect was also substantial because of the sheer size of the steel market—10,000 times that of the titanium market. China had been driving world steel consumption growth since the mid 1990s. Between 1999 and 2004, China's steel consumption doubled, accounting for about 39 percent of world consumption in 2004, more than the combined steel consumption of the United States and Japan. In 2007, China accounted for 31 percent

---

[22] In the process, DoD issued a waiver exempting all fasteners, a significant portion of the specialty metal business. The FY 2008 defense bill focuses more on specific exemptions than on the waiver process.

of world steel consumption and 34 percent of world steel production (IISI, undated).

The dramatic growth of China's steel consumption also drove up the price of vanadium, an alloy that is used in both steel production and titanium ingot production.[23] The vanadium PPI increased 360 percent during 2003–2005, when other raw materials for titanium production were also in extreme shortage. This also contributed to cost increases in titanium ingot production, since vanadium usually accounts for 4 percent of the weight of titanium ingot for aerospace use.

China may continue to be a wild card in the future titanium market. As we discuss in Chapter Six, China increased its titanium sponge production capacity dramatically—more than 300 percent in the two years between 2005 and 2007—and is expected to be the top or near-top titanium sponge producer in the world by 2010.

Summary

On the supply side, prices of titanium sponge and scrap started increasing sharply even before the commercial aircraft order surge in 2005 and 2006. Scrap was in extreme shortage in 2003–2005 due to the low airplane production rate, which resulted in less recycled scrap. The heightened demand for ferrotitanium, titanium scrap, and titanium sponge from carbon and stainless steel production further drove the shortage of titanium raw materials. The ferrotitanium PPI increased 82 percent from 2003 to 2005. This rapid price increase significantly influenced the prices of substitutes for ferrotitanium—titanium scrap and sponge. This cross-market substitution effect exacerbated the shortage of titanium raw materials, given the fact that the steel market is 10,000 times as large as the titanium market.

Titanium sponge prices also spiked because production could not keep up with demand. World titanium sponge production capacity declined 22 percent between 1997 and 2004, and U.S. sponge produc-

---

[23] The steel industry is one of the largest consumers of vanadium in the form of ferrovanadium (FeV). Vanadium increases the strength and toughness of steel.

tion capacity dropped 70 percent between 1995 and 2004. Given the long lead time necessary for adding production capacity and the severe downturn of the previous decade, producers were not able to respond quickly to the "unexpected" demand surge, driving up sponge prices.

The DLA titanium sponge stockpile depletion in 2005 coincided with the shortage in sponge and scrap. The depletion significantly influenced the titanium market because it had been an important buffer to mitigate titanium raw material supply fluctuations. Until 2005, titanium producers had been aggressively using the sponge stockpile as a substitute for scrap and ferrotitanium.

Since the titanium raw material supply was already tight in 2004, the additional demand shock from the record-high level of commercial aircraft orders in 2005 and 2006 further amplified the raw material shortage.

CHAPTER FIVE

Demand-Side Drivers of Titanium Price Fluctuations

Demand for titanium is determined by market conditions in the industries that buy titanium—so-called downstream industries. The aircraft manufacturing industry is the dominant buyer of titanium, and aircraft demand is highly cyclical. Other buyers are diversified among different industrial sectors and their demand for titanium moves with the economic growth rate, which is much less volatile than aircraft orders and deliveries.[1] Therefore, industry experts believe that fluctuations in the titanium prices are mainly driven by aircraft manufacturing, especially commercial aircraft demand cycles.[2]

Does this common belief explain the unprecedented price surge in the titanium market in recent years? What are the demand-side drivers of titanium price fluctuations? How did they contribute to the price surge? In this chapter, we explore demand-side price determinants, their changes in recent years, and how they are related to the dramatic price increase.

---

[1] Although the non-aerospace industrial sector as a whole accounts for more than half of the world's titanium consumption, the sector is highly segmented. A single part of the industrial equipment sector does not influence titanium prices anywhere near the extent to which aircraft manufacturing industry does.

[2] See Gambogi, 1998, and TIMET, 2005, 2006, among others.

Three Primary Demand Drivers of the Commercial Aircraft Manufacturing Industry

There have been three main demand drivers in the aircraft manufacturing industry in recent years. First, commercial aircraft orders skyrocketed as both Boeing and Airbus received a record number of orders during 2005 and 2006, as shown in Figure 5.1. Second, the titanium content per aircraft rose significantly, so that increases in aircraft orders in turn amplified the demand for titanium. Coinciding with the record-breaking increase in titanium demand from the commercial aircraft industry, the demand for titanium for military aircraft also increased significantly as full-time production of the F-22A Raptor began in 2003.

Commercial Aircraft Orders Skyrocketed

Boeing and Airbus collectively received orders for 2,139 aircraft in 2005 and 1,882 in 2006. These levels were almost twice that of previ-

Figure 5.1
Commercial Aircraft Orders and Deliveries, 1974–2006

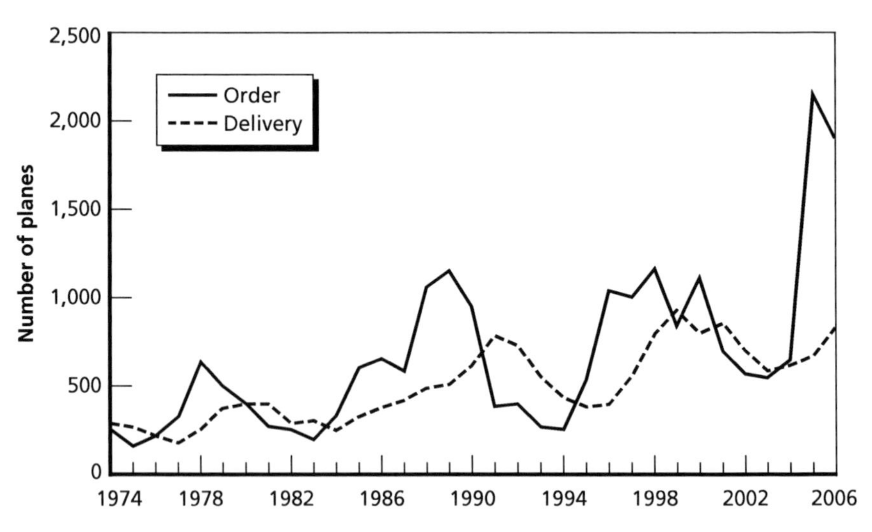

SOURCES: Boeing and Airbus company Web sites.
RAND *MG789-5.1*

ous peaks. Industry experts attributed this unexpected surge to growing air-traffic demands driven by an increase in passenger miles and the growing middle class in developing areas such as China and the Middle East (Rupert, 2006).

This historically high level of commercial aircraft orders coincided with the titanium price surge in 2005–2006. The sudden increase in commercial aircraft orders appears to be an obvious driver of the recent price surge. However, it is not clear whether the unusually large price rise in the titanium market in recent years can be fully explained by this demand-side surge. Deliveries of ordered aircraft are distributed over several years, so the delivery of titanium also is spread out. In other words, actual consumption of titanium has increased more slowly than the dramatic increase in aircraft orders may indicate, as depicted in Figure 5.1.

Titanium Content per Aircraft Increased

Average titanium material buy weight per commercial aircraft has increased substantially over the last 20 years, as shown in Figure 5.2.[3] Titanium MBW per aircraft increased from 10 tons for the Boeing 717, to 41 tons for the Boeing 747, to 91 tons for the Boeing 787. Given the demand for fuel-efficient commercial aircraft, new aircraft designs tend to have wider bodies and use more titanium and composites to produce lighter, cost-efficient aircraft. The compatibility between composites and titanium also has increased the use of titanium. The rising titanium content per aircraft amplified the impact of the increase in commercial aircraft orders in 2005 and 2006.

Increased Demand from Military Aircraft Manufacturers

Demand from military aircraft manufacturers has also increased significantly in recent years. Titanium MBW for military aircraft deliver-

[3] The average titanium MBW per aircraft is calculated as the ratio of total titanium MBW over all types of commercial aircrafts ordered each year to the number of plane orders in that year. The same calculation was done on delivery base. Data for the number of planes ordered and delivered were downloaded from the Boeing and Airbus Web sites. Titanium buy weights were obtained from Holz, 2006; TIMET; and RMI, 1994. Commercial aircraft included in the calculation are listed in Appendix A.

Figure 5.2
Average Titanium Buy Weight per Commercial Aircraft, 1984–2006

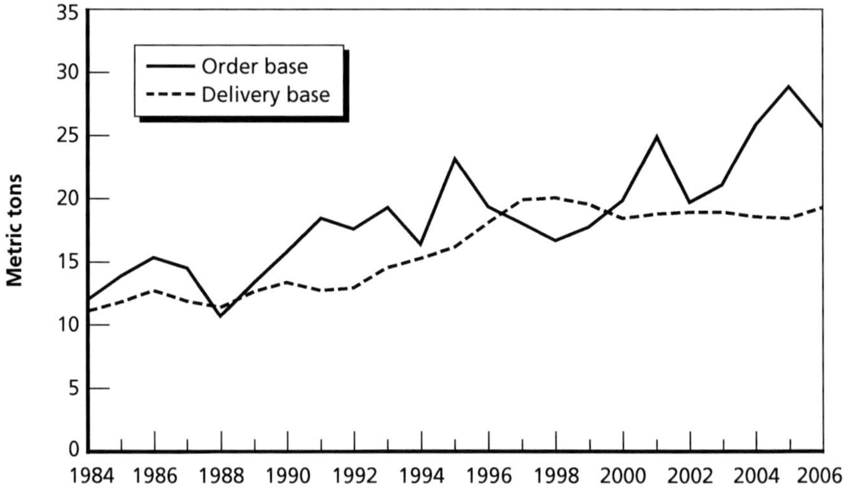

SOURCE: Authors' calculations based on the information summarized in Appendix A.
RAND *MG789-5.2*

ies increased 86 percent from 2000 to 2006, from about 1,380 tons to 2,560 tons (Figure 5.3). The U.S. military aerospace demand for titanium relative to the commercial demand from Boeing and Airbus—calculated from total titanium MBW for aircraft deliveries—increased from 9 percent in 2000 to 20 percent in 2003 and remained at that level until commercial aircraft orders surged in 2005.

F-22A deliveries accounted for about half of all U.S. military aircraft demand between 2003 and 2006. Titanium MBW per F-22A is about 50 tons, and 21 to 25 planes were delivered each year between 2003 and 2006.

The shortage of titanium since 2004 may have been exacerbated by this overlap of surges in military and commercial demand. As shown in Figure 5.4, military aircraft demand has generally moved together with commercial aircraft demand over the past decade, except between 2001 and 2003.

In addition, the F-35 program, one of the largest U.S. defense acquisition programs ever, is completing its development stage and is

Figure 5.3
Military Aircraft Titanium Buy Weight Based on Delivery Year, 2000–2006

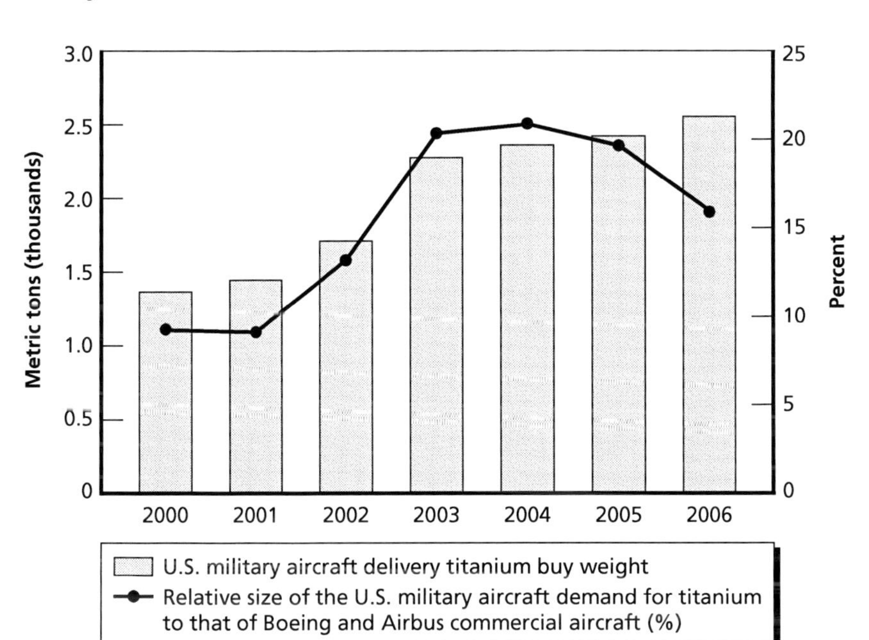

SOURCE: Authors' calculations based on the data sources listed in Appendix A.
NOTE: Military aircraft included in the calculation are listed in Appendix A.
RAND *MG789-5.3*

expected to increase the future titanium demand significantly when it goes into high-rate production. This may have influenced expected demand in the near future, which, in turn, may have influenced the current purchasing decisions of titanium buyers and thus the prices of titanium.

Increased Demand from the Industrial Sector

Before the surge in aircraft demand, the global titanium market was already tight because of high demand from industrial equipment manufacturers, the steel industry, and other titanium users.

Figure 5.4
Titanium Demand from Military and Commercial Aircraft Deliveries, 1991–2006

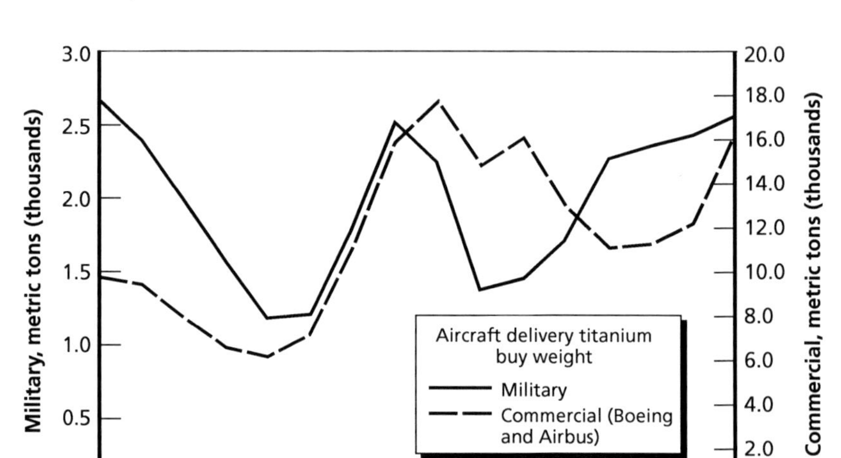

SOURCE: Authors' calculations based on the data sources listed in Appendix A.
RAND *MG789-5.4*

At the macro level, the strong economic growth of developing economies—for example, those of China and the Middle East—particularly since 2003, has created demand in industries such as chemical processing, power plants, and infrastructure construction, which need titanium-intensive equipment and materials. World steel production reached an historic high in 2004, driving up the price of alloys used in steel production—vanadium, ferrotitanium, and their substitutes.

The oil industry was among the emerging or growing sectors that require titanium and contributed to the growing titanium demand. The rising oil and gas prices in recent years induced growing demand for oil and gas production from deep-water sources, which requires extraction using titanium-intensive equipment.

In sum, the global industrial demand for titanium increased from 30,000 metric tons in 2003 to almost 42,000 metric tons in 2006—about a 40 percent growth over the three-year period, as shown in

Figure 5.5.[4] During the five years between 2001 and 2006, global titanium shipments to the industrial sector grew faster than those to the aerospace industry—53 percent and 21 percent, respectively. In addition, the industrial sector demand for titanium began an upswing in 2001, before the increased demand from the aerospace sector.

Increased Spot Market Transactions

In recent years, titanium price volatility also has been driven by the increase in spot market transactions. The reasons for the increase are rooted in the unexpected nature of the recent demand surge.

First, buyers with long-term contracts still depended on the spot market, because the quantity they secured with long-term contracts

Figure 5.5
Global Titanium Demand by Sector, 1997–2006

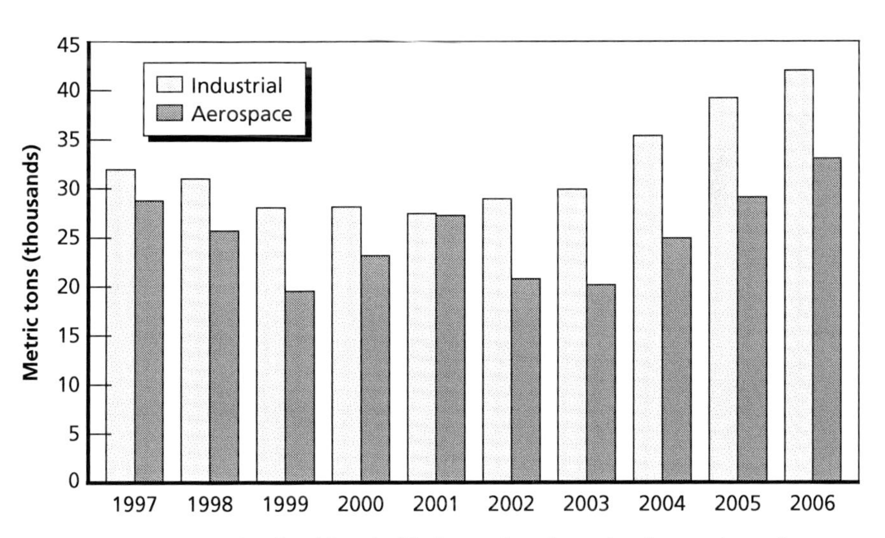

SOURCES: TIMET, 2006; other historical industry data from titanium melters, forgers, and casters.
RAND *MG789-5.5*

---

[4] The industrial sector includes all industries except the aerospace industry. The aerospace industry includes both the commercial and military sectors.

was less than they wanted to buy during 2005 and 2006. In addition, the supply shortage created longer lead times for titanium deliveries, forcing buyers to use the spot market despite its much higher prices.[5] Second, during the previous downturn of the titanium industry, many buyers—including aircraft manufacturing subcontractors—preferred spot market transactions over long-term contracts because spot prices were cheaper and availability was not a problem. Third, as the supply shortage became serious, speculative buying and hoarding naturally occurred, which in turn increased spot transactions.

Historically, the titanium spot market has accounted for a relatively small proportion of the titanium purchased by the aircraft manufacturing industry. Aircraft manufacturers usually procure titanium through five- to ten-year long-term contracts based on projected aircraft build rates. However, in 2005 and 2006, supply shortages forced even aircraft manufacturers to procure some portion of their titanium metal on the spot market.[6]

As happened during the previous titanium metal shortage in the early 1990s, there was also some panic and speculative demand for titanium sponge, ingot, and mill products to secure future supplies, although the extent of hoarding is unknown.

In this strong seller's market, titanium prices were subject to the bargaining power of the supplier, who had significant leverage in relation to the buyer. Titanium producers had a golden opportunity to

---

[5] The titanium delivery lead time has increased by up to three times. According to a major titanium buyer, there were cases in which titanium producers quoted neither price nor delivery time.

[6] The long-term contracts by aircraft manufacturers such as Boeing and Airbus with titanium producers were also handled loosely. The long-term contracts were made on behalf of parts subcontractors to Boeing and Airbus, but the subcontractors, knowing that they could pay lower prices on the spot market, did not want to participate in those contracts during the 1998–2003 downturn in the market. This led to the use of the spot market by the subcontractors, and in some cases to legal disputes between the titanium producers and aircraft manufacturers for breach of long-term contracts. When the market took an unexpected upturn in 2005, this practice backfired on many subcontractors who then were exposed to short-term price volatility. Some military aircraft subcontractors had an even greater exposure to the spot market because they had to purchase titanium for one lot production at a time.

recover their losses from the severe downturn between 1998 and 2003, right before the demand surge in 2004.[7]

Interaction of Demand- and Supply-Side Drivers to Bring Out the Recent Turmoil in the Titanium Market

How the price drivers from both sides unfolded over time to trigger the recent titanium market turmoil is detailed in Appendix C.

In sum, increased demand for titanium beginning in 2004 exceeded the available supply of scrap and sponge and also exceeded the production capacity for new titanium metal. Given the high capital investment and long lead times required for the expansion of titanium sponge production capacity, sponge supply expansion was not responsive enough to meet the unexpected surge in demand during the short run. Moreover, given the long record of excess capacity in the industry, titanium producers were reluctant to invest in capacity expansion until they were assured the strong demand would persist for at least several years. With increased spot market transactions and speculative demand, titanium prices skyrocketed and titanium spot market prices became extremely volatile.

Relationship Between Titanium Price Trends and Demand Shocks from the Aircraft Manufacturing Industry

According to industry experts, cyclical fluctuations of titanium prices are mainly driven by demand-side events, especially aircraft demand cycles (Gambogi, 1998; TIMET, 2006).

Table 5.1 lists significant events in the titanium market between 1971 and 2005 compiled from the literature. The list is dominated by demand-side events, especially commercial aircraft industry business

---

[7] Even in the summer of 2004, major U.S. titanium producers were laying off workers to cut losses.

cycles. Of the 16 events listed, 13 are from the demand side and nine of those are from the commercial aircraft manufacturing industry.[8]

Commercial aircraft orders and deliveries are highly cyclical, as depicted previously in Figure 5.1. Commercial aircraft orders skyrocketed in 2005 and 2006, corresponding with the price surge after 2004. The sudden increase in commercial aircraft orders seems to be an obvious driver of the recent price surge. However, it is not clear whether the unusual rate of titanium price inflation in recent years can be fully explained by this demand-side event in the commercial aircraft industry.

We matched the PPI trend with the major events of the industry listed in Table 5.1, as shown in Figure 5.6. Interestingly, demand shocks from the commercial aircraft manufacturing industry appear to correlate with price movements until 1997 and after 2004. However, the price trend during the titanium market downturn from 1998 to 2003 does not seem to match the cyclical fluctuations of commercial aircraft industry demand.

Titanium Demand from the Commercial Aircraft Industry and Titanium Price Trends

To examine the sensitivity of titanium mill product prices in relation to the titanium demand growth that was driven by shocks in the aircraft manufacturing industry, we matched the quantity of the commercial airline industry's demand for titanium with the mill product PPI trend. We calculated titanium demand by using the number of commercial aircraft deliveries per year since the mid 1980s.[9] We assumed titanium

---

[8] Our study identifies the significant events of the industry in a more balanced manner than existing studies, which assume that aerospace demand drives titanium market fluctuations. Appendix C summarizes how significant events on the supply and demand sides unfolded over time to trigger the recent titanium market turmoil.

[9] We used aircraft delivery data in the demand calculation because we could not obtain titanium purchase data. From the plots of titanium demand based on deliveries and PPI price trends, we observe the responsiveness of the titanium price to aircraft deliveries.

Table 5.1
Significant Events Affecting the Titanium Market, 1971–2005

Year(s)	Event	Supply	Demand	Civilian Aircraft Demand	Military Aircraft Demand	Event Identity Code[a]
1971	Research for supersonic transport terminated		X	X		
1975–1976	Military aircraft production peak (F-14 and F-15)		X		X	A
1977–1981	Rapid increase in orders for commercial aircraft		X	X		B
1982–1984	Collapse of the commercial aircraft market		X	X		C
1984–1986	Production of B1-B bombers		X		X	D
1985–1989	Renewed strength in the commercial aircraft market		X	X		E
1988–1989	Increases in U.S. sponge production capacity	X				F
1990–1994	Reductions in military and commercial aerospace demand		X	X	X	G
1992	Sodium-reduction sponge plant closed at Ashtabula, Ohio	X				H

Table 5.1—Continued

Year(s)	Event	Supply	Demand	Civilian Aircraft Demand	Military Aircraft Demand	Event Identity Code[a]
1993	Magnesium-reduction sponge plant commissioned at Henderson, Nev.	X				I
1994–1997	Surge in consumer goods and commercial aerospace orders		X	X		J
1997–1998	Cancellation of some commercial aircraft orders		X	X		K
1999	Initial production of the F-22 starts		X		X	L
2001	Decline in the commercial airline industry		X	X		M
2003	Full-time production of the F-22A		X		X	N
2005	Commercial aircraft orders skyrocket		X	X		O

SOURCES: Gambogi, 1998; ODUSD-IP, 2005; TIMET, 2006.

[a] The event identity code relates to the timeline in Figure 5.6.

Figure 5.6
PPI Fluctuations for Titanium Mill Shapes and Supply and Demand Shocks in the Industry

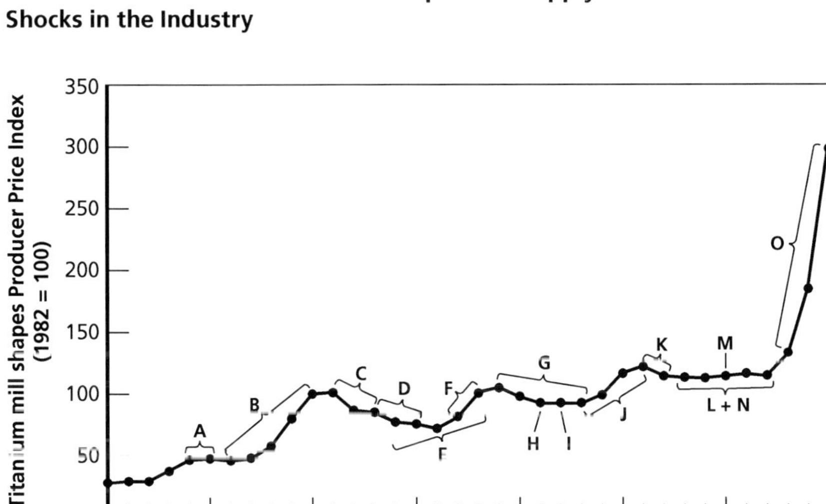

SOURCE: PPI data from the Bureau of Labor Statistics.
RAND *MG789-5.6*

is delivered approximately one year before the aircraft.[10] For a given year (t), titanium demand from each type of aircraft was calculated as the titanium MBW for the aircraft multiplied by the number of planes delivered in year (t + 1).[11] The names of the aircraft included in the demand calculation and the sources of data are listed in Appendix A.

A few interesting characteristics can be observed by comparing the demand from commercial aircraft delivery demand and the PPI trend of titanium mill shapes in Figure 5.7. First, the aircraft indus-

---

[10] Our assumption of a one-year lead time is based on the TIMET 2006 annual report. The actual time lag between titanium delivery and delivery of aircraft would vary for different titanium products. According to experts from the aircraft manufacturing industry, the lead time varies from 30 days to 18 months.

[11] MBW is a measure of how much titanium should be purchased to manufacture each aircraft. MBW is usually much larger than MFW, which is a measure of how much titanium is actually included in the finished aircraft. The difference between the MBW and MFW is due to the scrap generated from the manufacturing process. MBW includes titanium required to be purchased for both airframes and engines.

Figure 5.7
Titanium Demand from Commercial Aircraft Deliveries and Titanium Mill Shapes PPI Trend, 1985–2005

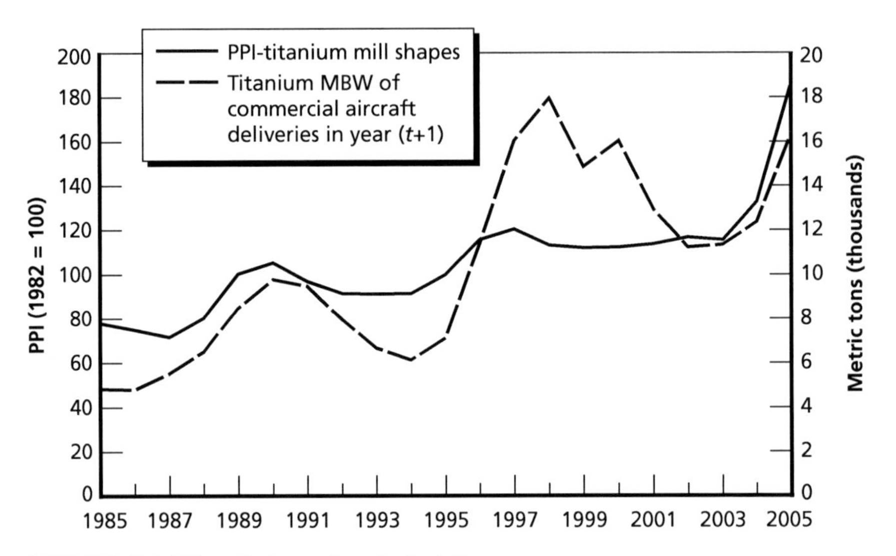

SOURCES: BLS, PPI statistics; authors' calculations.
RAND *MG789-5.7*

try titanium demand and titanium PPI are correlated in most years, except between 1998 and 2003.[12] This corresponds with the observation above in Figure 5.6. Second, the titanium PPI is less volatile than the quantity of titanium needed to meet aircraft deliveries, except in 2004 and 2005. Unlike in other periods, titanium prices increased faster than titanium demand in 2004 and 2005. While the aircraft industry's demand grew 9 percent and 31 percent in 2004 and 2005, respectively, titanium PPI price growth rates were 15 percent and 39 percent, respectively, in those years.

Titanium Mill Product Price Elasticity Before 2004

The PPI for titanium peaked in 1997 as mill product shipments peaked, as shown in Figure 5.8. After that, even as the demand for titanium by the commercial aircraft industry declined from 1998 to 2003, the

[12] The Pearson correlation coefficient is 0.77.

PPI index for titanium mill shapes remained quite stable.[13] In other words, contrary to the common belief that the titanium PPI is driven largely by demand from the commercial aircraft industry, the titanium mill product price trend from 1997 to 2003 seemed less sensitive to the declining demand from the commercial aircraft industry than during other periods.

Why was the titanium mill product PPI relatively insensitive to the decline in titanium demand from the commercial aircraft industry during 1998–2003? Figure 5.8 shows how the mill product shipments of U.S. producers moved with the aerospace demand for titanium, as both declined significantly during that period. This linkage reflects the dominance of aerospace buyers in the U.S. titanium market.

Figure 5.8
Trends of U.S. Titanium Shipments, Demand from Commercial Aircraft Deliveries, and PPI, 1996–2005

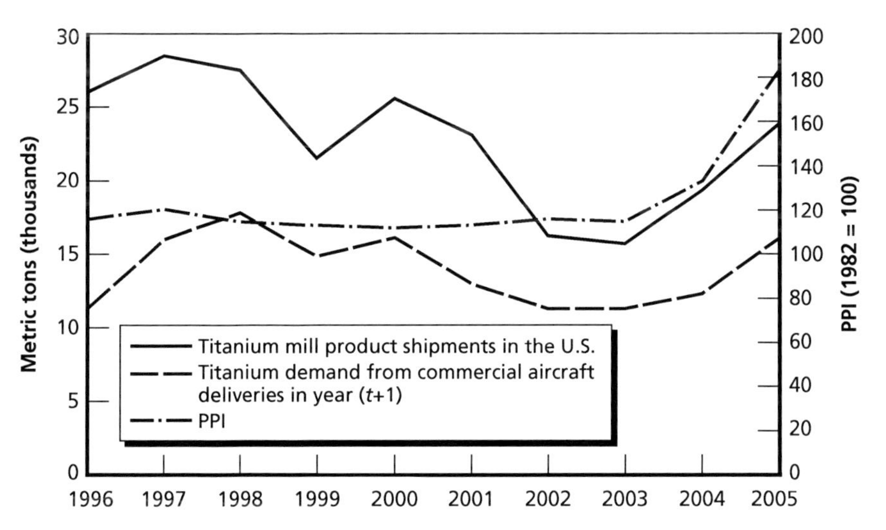

SOURCES: USGS Mineral Industry Surveys, 2006; BLS, PPI statistics; authors' calculations.

RAND *MG789-5.8*

---

13 Prices of titanium raw materials such as titanium sponge, scrap, and ingot are more volatile than the mill product PPI. (Titanium raw material price trends were discussed in the previous chapter.)

However, during the same period, world titanium shipments did not decrease as sharply as shipments by U.S. producers. As shown in Figure 5.9, U.S. shipments of titanium mill products in 2003 constituted only 56 percent of the previous peak in 1997, whereas world shipments were about 83 percent of their 1997 peak in 2003 and had fully recovered by 2004.

Similar to other raw material markets, the titanium market is globalized and domestic prices of titanium move together with those in the rest of the world. The relative insensitivity of the titanium PPI to commercial aerospace's demand for titanium between 1998 and 2003 may reflect the fairly stable global demand for titanium, in contrast with the volatile aerospace demand. In the global titanium market, industrial demand, historically more stable than aerospace demand, has dominated aerospace demand since the mid 1990s, as observed in

Figure 5.9
Titanium Mill Product Shipment Trend in the United States Compared with That in the Rest of the World

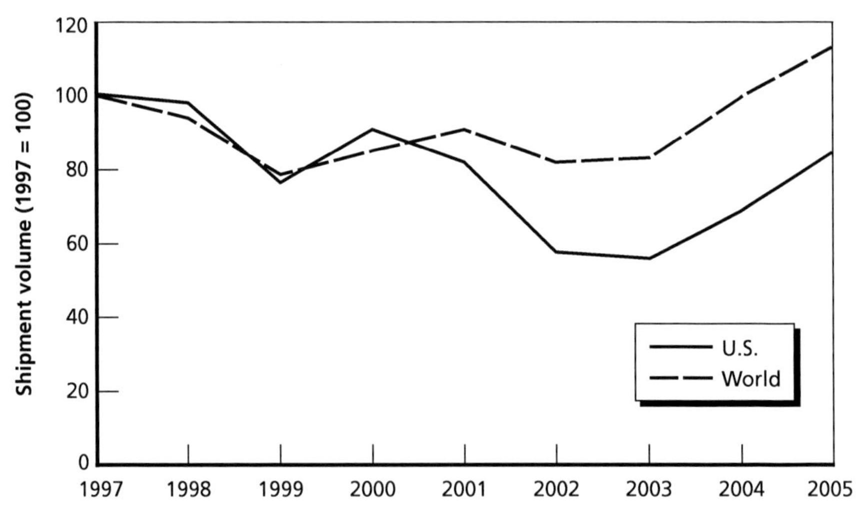

SOURCES: USGS *Minerals Yearbook*, various years; TIMET, 2006; data from titanium melters, forgers, and casters.
NOTE: Shipment volume in metric ton for each year is normalized by the tonnage in 1997.
RAND *MG789-5.9*

Figure 5.5. In addition, the industrial titanium market bottomed out in 2001, two years earlier than the aerospace market.

Long-term contract practices may also have contributed to titanium price stability between 1997 and 2003.[14] Even though year-to-year titanium demand from commercial aircraft deliveries fluctuated from 1998 to 2003, the annual average demand for that period was approximately 12 percent lower than the rate in 1997. The decline in commercial aircraft demand for titanium may have been evenly distributed over the six-year period through long-term contracts, instead of being reflected in year-to-year price changes.

While the titanium PPI reflects market price trends for all segments of the titanium market, prices of titanium mill shapes for the aircraft manufacturing market segment may have been more responsive to the decreased demand between 1998 and 2003. In fact, aircraft industry experts mentioned that short-term prices of titanium mill products in this period were significantly lower than the 1997 peak and that the PPI for 1998 to 2003 did not seem to represent the actual prices paid by aircraft manufacturers.[15]

Isolated titanium price data for the aerospace segment are not available. Aerospace is TIMET's dominant customer, so TIMET's average mill product prices are a close substitute. As Figure 5.10 shows, while TIMET's average mill product prices moved around the PPI trends, its prices fluctuated more than the PPI from 1998 to 2003. TIMET's lowest average price during the period was 18 percent lower than the peak price, while the PPI fluctuation was less than 8 percent lower than its peak.

However, this comparison does not change the conclusion that between 1998 and 2003, overall titanium mill product prices were less sensitive to the declining demand from commercial aircraft manufacturing than they were in other periods.

---

[14] The PPI of titanium mill shapes is based on both spot market prices and long-term contract prices, similar to the universe of transactions in the titanium producer marketplace.

[15] Authors' discussion with Boeing representatives in July 2007.

Figure 5.10
Comparison Between the PPI for Titanium Mill Shapes and TIMET's Average Mill Product Price, 1996–2004

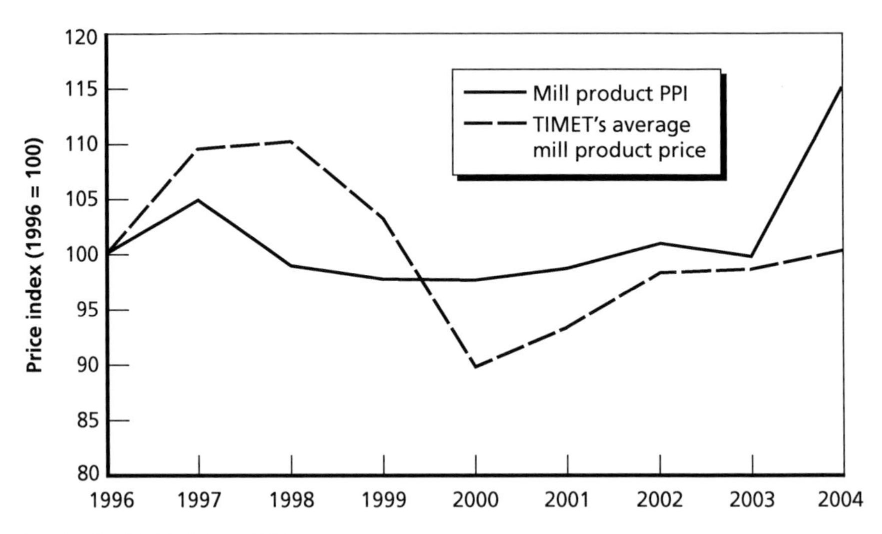

SOURCES: BLS; TIMET, 2006.
RAND *MG789-5.10*

Price Elasticity of Titanium Demand Since 2004

Why did titanium prices increase faster than demand growth in 2004 and 2005, unlike in previous periods? There are several possible explanations for the greater price volatility in recent years.

First, titanium orders increased much more sharply than the actual consumption of titanium in 2005 and 2006. This unexpected demand surge in aircraft orders is well illustrated by the *Airline Monitor* forecast of commercial aircraft deliveries issued in January 2004, which forecasted that commercial aircraft deliveries would not start increasing again until after 2007.[16] It was not until July 2004 that *The Airline Monitor* forecasted a sharp upturn in the aircraft manufacturing industry. According to industry experts, U.S. titanium producers were

[16] *The Airline Monitor* is a leading publication that publishes forecasts for commercial aircraft deliveries for the next 20 years or so. The forecasts come out in January and July each year.

laying off their employees in some of their factories as late as summer 2004, unaware of the impending surge in aircraft demand.[17]

To the extent that the dramatic increase in titanium demand was unexpected, the supply of titanium was not responsive enough.[18] As a result, there were large order backlogs and longer lead times in the delivery of titanium products. Speculative buying and possible hoarding clouded the demand picture, adding spot market transactions that further increased price volatility. Therefore, the market price in 2004–2005 may have reflected not only the actual demand but also the amplified expected demand for the future.

On the other hand, the dramatic movement in titanium prices in recent years was also triggered by factors other than the demand surge in the commercial aircraft industry. Titanium prices were influenced by supply-side drivers, such as raw material availability and price movement of substitute metals. In fact, price movements of both ferrous and nonferrous metals became generally more volatile after 2003, as shown in Figure 5.11. In particular, the nonferrous sector has continued on a volatile price trend even as ferrous metal prices have settled.

Summary

Industry experts believe that cyclical fluctuations of titanium prices are driven mainly by demand-side events, especially aircraft demand cycles. However, during the previous titanium market downturn (1998–2003), the PPI for titanium mill shapes in the United States was relatively insensitive to the declining demand from the commercial aircraft industry, contrary to expectations.

Titanium price insensitivity to commercial aircraft demand in 1998–2003 was partly due to the dominance of non-aerospace industrial demand and the strong growth of that sector. In the global titanium market, industrial demand, historically more stable than aerospace demand, had dominated aerospace demand since the mid 1990s.

---

[17] Authors' discussion with TIMET in August 2007.

[18] We discussed supply responsiveness in Chapter Four.

Figure 5.11
PPI Trends for Various Metals, 1986–2006

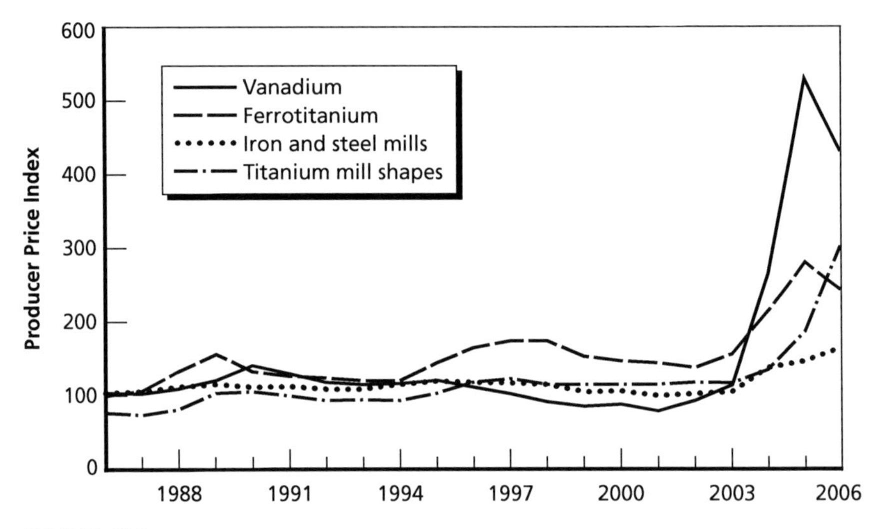

SOURCE: BLS.
RAND *MG789-5.11*

The industrial titanium market bottomed out in 2001, two years earlier than the aerospace market. By 2004, however, world titanium demand was already at its previous peak level, driven by strong growth in the non-aerospace industrial sector. This amplified the impact of the commercial aircraft order surge on titanium prices in 2005 and 2006.

Recent skyrocketing prices and extreme supply shortages in the market were triggered by the coincidence of a series of events generated from both the supply and demand sides.

On the demand side, the sudden increase in commercial aircraft orders in 2005–2006 was one of the obvious price drivers. The trend toward greater titanium content per aircraft amplified the effect of the increased aircraft manufacturing demand on titanium prices. In addition, increases in military and industrial demand for titanium coincided with the demand surge from the commercial aircraft industry.

Titanium price volatility was further exacerbated by the increase in spot transactions on the titanium market during 2005–2006. During that period, even aircraft manufacturers had to procure titanium on the spot market due to the supply shortage and long lead times. During the strong seller's market, titanium prices were subject to the titanium producer's bargaining power. Titanium producers had a golden opportunity to recover from the severe five-year downturn that immediately preceded the demand surge since 2004.

On the supply side, increased demand for titanium beginning in 2004 exceeded the available supply of scrap and sponge and also exceeded the production capacity for new titanium metal. Given the high capital investment and long lead times required for the expansion of titanium sponge production capacity, sponge supply expansion was not responsive enough to meet the unexpected surge in demand during the short run. Moreover, given the long record of excess capacity in the industry, titanium producers were reluctant to invest in capacity expansion until they were assured the strong demand would persist for at least several years. With increased spot market transactions and speculative demand, titanium prices skyrocketed and titanium spot market prices were extremely volatile.

CHAPTER SIX

Market Prospects and Emerging Technologies

In previous chapters, we examined what triggered the recent titanium price surge by analyzing historical data and events in the industry. In this chapter, we review future market prospects and emerging technologies and their cost-saving potential. However, a rigorous forecasting of the future titanium market and technological improvements is beyond the scope of this study. Rather, the purpose of this chapter is to develop a better sense of possible titanium market conditions and technological developments in the near future.[1]

Market Prospects

Prospects of the World Titanium Sponge Supply

In response to the recent demand surge, many titanium metal producers announced or took steps to increase sponge capacity in the near future.[2] China has been the most aggressive in expanding titanium

---

[1] Our inferences are based mainly on our interviews with industry experts, our understanding of the industry, and related literature.

[2] Future sponge capacity expansion statistics for United States, Japan, and Russia are compiled from annual reports and press releases of major sponge producers in those countries, including TIMET, ATI, RTI, VSMPO, Toho Titanium (Mitsui & Co., Ltd.), and Sumitomo Titanium Corporation, as of February 2008. Data for China, Kazakhstan, and Ukraine are cited from the USGS Mineral Commodities Summary: Titanium and Titanium Dioxide, 2008. The USGS statistics reflect capacity expansion plans announced by 2007. The 2007 production capacity statistics are estimated levels, not actual.

sponge production capacity. Chinese sponge producers expanded their sponge capacity dramatically from 9,500 tons per year in 2005 to 45,000 tons per year by 2007, an increase of approximately 373 percent in two years (Figure 6.1). They plan to achieve a capacity of 50,000 tons per year by the end of 2008. Japanese sponge producers have a capacity expansion plan of 52,000 tons per year by 2009, a 40 percent increase from its 2005 capacity level of 37,000 tons per year. Russian sponge capacity is expected to increase to 44,000 tons per year by 2011.

In the United States, titanium producers increased their sponge production capacity dramatically from 8,940 tons per year in 2005 to 20,200 tons per year in 2007 and plan to achieve a historical high level of 41,970 tons per year by 2010. Thanks to the recent sharp increase in titanium prices, many titanium producers hold a strong cash position and now are able to afford large-scale capital investment plans. As a result, some of the U.S. sponge capacity expansion plans will be funded primarily by cash on hand as well as operating cash flow (RTI International Metals, Inc., 2007a).

TIMET, the largest sponge producer in the United States, is expected to increase the capacity of its sponge plant in Henderson, Nevada, to 12,600 tons per year by 2008. In early 2007, TIMET planned to select a site where it would have built a new sponge plant with an initial capacity of 10,000 to 20,000 metric tons so that it could achieve a sponge capacity of at least 22,600 tons per year (and as much as 32,600) by 2010. However, this plan was canceled in November 2007. Instead, TIMET entered into a long-term sponge purchase agreement with Toho Titanium.

Allvac (ATI) restarted its idle sponge plant in Albany, Oregon, in 2006, and it is expected to reach its capacity of 7,260 tons per year by the end of 2007. By the end of 2008, ATI plans to expand the capacity of its Oregon plant by 1,800 tons and to open a new sponge plant in Rowley, Utah, with a capacity of 10,900 tons per year. In total, ATI is expected to expand its capacity to 19,960 tons per year by 2009.

In September 2007, RTI announced plans to build a new sponge factory with a capacity of 9,070 tons per year by 2010. RTI also plans

Figure 6.1
Planned Expansion of World Titanium Sponge Capacity Through 2010

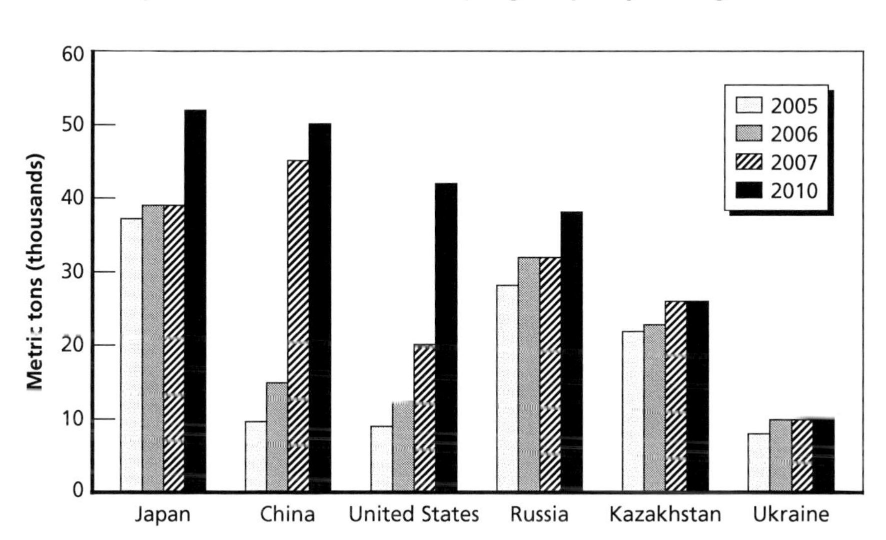

SOURCES: Capacity for the major titanium sponge companies from the United States, Japan, and Russia was compiled from news releases and annual reports of the companies. Data for other countries were obtained from USGS, Mineral Commodities Summary, 2008.
NOTES: We assumed that sponge capacity of Kazakhstan and Ukraine in 2010 is the same as that in 2007, because information on their future capacity plans is not available. Capacity data for the United States, Japan, and Russia have been updated as of February 2008. Other country data reflect plans announced by 2007. Russian capacity in 2010 is assumed to reach the midpoint between its 2007 and 2011 capacities. Chinese capacity in 2010 is assumed to be the same as its 2008 capacity. We included four years (2005, 2006, 2007, and 2010) because we do not have data for each year after 2007.
RAND *MG789-6.1*

to double its mill product production capacity from 7,500 tons in 2007 to 15,000 tons in 2010.

Given the expansion plans detailed above, the world titanium sponge capacity is estimated to almost double from 113,540 tons per year in 2005 to 217,970 tons per year in 2010. If the new plants are in full operation as pictured in Figure 6.1, China is expected to be the second-largest titanium sponge producer in the world by 2010. Because the projected difference in sponge capacity between Japan and China are based on a conservative assumption about

Chinese expansion, we cannot exclude the possibility that China may take over the first place in sponge production by 2010. In terms of expansion rate between 2005 and 2010, China is the most aggressive player, followed by the United States. According to the current capacity expansion plans of U.S. producers, the U.S. sponge capacity in 2010 will be 4.7 times that of 2005. Once this expansion has been completed, this increase in titanium sponge capacity will make the United States the world's third- or fourth-largest (depending on the speed of Russia's capacity expansion) sponge producer, led by Japan and China.

The successful completion of the American and Chinese expansion plans is one of the key concerns of the titanium industry. These aggressive expansion plans may help the titanium market reach equilibrium if an optimistic demand forecast comes true in the future. However, they could contribute to a significant market imbalance if future demand does not meet expectations.

The Impact of China on the Titanium Supply

Chinese demand and the expansion of its production capacity of various metals has been a wild card in the titanium supply.[3] This situation may continue in the near future because of the sheer size of China's economy and its remarkably fast, seemingly ceaseless, growth. In 2003–2005, the Chinese steel industry consumed considerable amounts of titanium scrap and sponge, which drove up market prices for both commodities and provided an incentive for the construction of a multitude of sponge plants in China. However, the combination of slightly reduced growth in China's steel demand and the significant increase in sponge production capacity had the reverse effect on the market—spot market prices of sponge dropped nearly 50 percent in 2007. Responding to the market situation, China's capacity expansion efforts are expected to slow down considerably after 2007. However, no one can exclude the possibility that China will restart its aggressive capacity expansion effort at any time in the near future.

---

[3] China's rapid economic growth has also contributed to growth in the industrial demand for titanium, which is the largest consumption sector in the world titanium market. However, China's direct consumption of titanium is still relatively small.

Future Demand for Titanium

Industrial Demand. From 2007 to 2010, the industrial demand for titanium is expected to increase faster than the world GDP growth rate. This increased demand will be driven largely by the growth in additional titanium applications throughout the industrial sector. For example, demand growth among emerging titanium buyers, such as oil and gas installations and the automobile, heavy vehicle, and medical device industries, has been particularly dramatic in recent years—some 50 percent between 2004 and 2006. Thanks to this dramatic growth, titanium demand from the industrial sector as a whole increased 19 percent during 2004–2006.

Looking ahead, even as the growth rate in industrial applications for titanium is forecasted to slow down, many experts expect the industrial sector demand for titanium to grow 14–15 percent from 2007 to 2010, with a compounded annual average growth rate (CAGR) of about 4 percent for those years.

Commercial Aerospace Demand. Based on the record-breaking commercial aircraft orders for Boeing and Airbus in 2005 and 2006, titanium demand from the commercial aerospace sector is expected to grow rapidly in next several years, at least until 2010. According to *The Airline Monitor,* the number of commercial aircraft deliveries is forecasted to increase rapidly—at a CAGR of 9 percent between 2005 and 2010—and then slow until 2014 when aircraft deliveries are forecasted to increase sharply again.

Even if the growing aircraft delivery forecast is realized, the titanium demand trend is expected to grow much faster (22 percent CAGR) than the number of planes to be delivered because of the significant growth in titanium buy-weight per plane (Figure 6.2). This rapid growth in titanium demand is driven primarily by orders for large, titanium-intensive aircraft such as Boeing's 777 and 787, and the Airbus 380.

According to our calculation for Figure 6.2, these three titanium-intensive aircraft will account for more than half of the commercial aircraft titanium demand from Boeing and Airbus in 2010. The Boeing 787, in particular, requires about 91 metric tons of titanium per plane, and is expected to account for almost 30 percent of the Boeing and

Figure 6.2
Forecasted Commercial Aircraft Deliveries and Future Titanium Demand

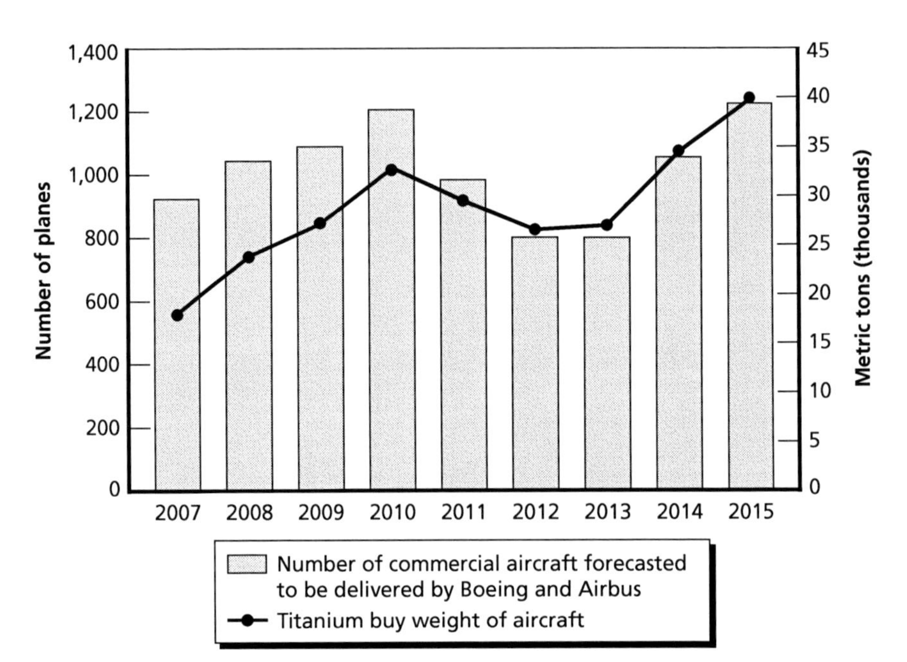

SOURCE: *The Airline Monitor*, 2007, for forecasted commercial aircraft deliveries.
NOTES: The titanium buy weight of commercial aircraft deliveries was calculated by summing the products of the number of planes to be delivered each year and the titanium buy weight per aircraft. The types of aircraft included in the calculation and the data source of titanium buy weight of each aircraft are summarized in Appendix A.
RAND *MG789-6.2*

Airbus titanium demand in 2010, assuming the *Airline Monitor* delivery forecast is correct.[4]

Military Demand. As worldwide defense spending continues to grow, military weapon development and acquisition continue to evolve in favor of lighter armaments with greater mobility. As a result, the military consumption of titanium will continue to increase in the future. TIMET's 2005 forecast expects that defense spending for all systems will remain strong until 2010. Current U.S. military aircraft programs

[4] To calculate future titanium demand, we used the *Airline Monitor*'s January 2007 forecast of aircraft deliveries.

such as the C-17 and F-15 are expected to continue production until 2010.[5] Production of the F/A-18 E/F and F-16 are expected to continue until 2015. Full-rate production for the F-22 Raptor began in 2003 and is expected to continue through 2009. The F-35 is expected to enter low-rate initial production in 2008, and the delivery of the first production aircraft is scheduled for 2010.[6] F-35 production levels could be as many as 3,500 planes over the next two decades, including sales to foreign countries.[7] However, U.S. military aircraft demand for titanium in 2010 is expected to be significantly lower than that in 2005 because the F-22 production will be completed and F-35 production will only be ramping up by 2010.

Assuming that commercial aircraft demand in 2005–2010 will be at 22 percent CAGR, as discussed above, the total aerospace demand for titanium—both commercial and military—is estimated to be 16 percent CAGR for 2005–2010.[8]

Future Scenarios of World Titanium Demand. What will world titanium demand look like in the near future, say 2010? To better evaluate future titanium demand, we developed three scenarios—base, optimistic, and pessimistic—based on various assumptions about growth rates in each market segment for the period 2005–2010, as detailed in Table 6.1. The assumptions we made are ad hoc; they reflect a diversity of opinions from industry experts who regularly assess demand prospects.[9]

---

[5] F-15 production is for exporting to foreign buyers.

[6] The Air Force funded production of the first two aircraft in fiscal year 2007. Refer to Defense Acquisition Management Information Retrieval Web site, Selected Acquisition Report (SAR) for F-35 Joint Strike Fighter, December 31, 2006.

[7] According to the SAR, DoD plans to procure about 2,000 F-35 planes until 2028.

[8] It is assumed that military aircraft demand for titanium will decrease by 20 percent CAGR in 2005–2010, while commercial aircraft demand will grow 22 percent CAGR in the same period, as discussed above. Some optimistic forecasters believe that total aerospace demand will be higher than 16 percent CAGR; some experts disagree and believe it will be much lower.

[9] The assumptions on growth rates of different market segments are calibrated from what industry experts envision as the level of total titanium demand in 2010. To obtain a reasonable set of assumptions, we repeated the calibration until we confirmed that the calibrated

The CAGR for each market segment assumed in Table 6.1 are higher than the currently forecasted growth rates for the period of 2007 and after, reflecting the growth spurt in industrial demand and the dramatic increase in aircraft orders in 2005 and 2006. In all three scenarios, we used 2005 as the base year in order to compare future demand situations with future production capacity.

Table 6.1 summarizes the future demand scenarios and estimated titanium demand in 2010 for each scenario. In the base scenario, we assume a CAGR for aerospace and industrial demand at 15 percent and 5 percent, respectively, during 2005–2010.[10] Under this scenario, aerospace demand doubles and industrial demand increases 28 percent during the five-year period, so that world titanium demand in 2010 becomes 59 percent higher than that in 2005. Aerospace's share in world titanium consumption reaches 54 percent in 2010 under the base scenario.

In the optimistic scenario, we assume 20 percent and 7 percent CAGR for aerospace and industrial demand, respectively, for 2005–

Table 6.1
Future Scenarios of World Titanium Demand in 2010

		Scenario		
	Demand Sector	Base	Optimistic	Pessimistic
Compounded annual average growth rate of each market segment (%)	Aerospace	15	20	10
	Industrial	5	7	3
Titanium demand in 2010 (base year 2005 demand = 100)	Aerospace	201	249	161
	Industrial	128	140	116
	Total demand	159	187	135

NOTE: The CAGR of each market segment in the table is calculated for 2005–2010, with 2005 as the base year.

results were fairly consistent with historical data for the level and composition of titanium demand, existing forecast of commercial aircraft demand, and industry experts' vision of 2010 titanium demand.

[10] We use CAGRs as qualifiers to distinguish different scenarios because they intuitively illustrate the speed of growth in each segment for the given period.

2010. Under this scenario, aerospace demand increases 149 percent and industrial demand increases 40 percent, causing total world titanium demand in 2010 to be 87 percent greater than that in 2005. In this scenario, aerospace's share of world titanium consumption will be 57 percent in 2010.

The pessimistic scenario assumes 10 percent and 3 percent CAGR, respectively, for aerospace and industrial demand in 2005–2010. Under this scenario, aerospace demand increases 61 percent and industrial demand increases 16 percent, resulting in a total world titanium demand that is 35 percent higher in 2010 than it was in 2005.

Which scenario is the most likely? It is not possible to determine the likelihood of each demand scenario. However, we can infer the strength of future demand from a few core indicators, such as forecasted commercial aircraft deliveries and titanium requirements for each aircraft, because commercial aerospace currently accounts for about half of the titanium demand and it is a main source of market fluctuations.

The optimistic scenario is more likely if the build rates for titanium-intensive aircraft, particularly the Boeing 787 Dreamliner and the Airbus 380, continue to increase. The pessimistic scenario is more likely if production of the Boeing 787 lags behind the planned schedule and Airbus continues to delay production of its titanium-intensive A350 and A380.[11]

Summary: Future Titanium Market Balance

Although we are not equipped to decisively predict the most likely outcome for the titanium market in the future, we have identified the factors that likely will influence the future titanium market balance.

---

[11] In September 2007, Airbus and RTI International Metals announced a contractual arrangement of $1.1 billion for 2010–2020 that includes A380 and A350 programs. However, it is not unusual for aircraft manufacturers to cancel their orders, especially if the orders are for delivery a long time in the future.

Important factors that could influence the outlook for the market include

1. the realization of capacity expansion plans by titanium suppliers, particularly American and Chinese producers
2. the Boeing 787 build rate and demand from other titanium-intensive emerging aircraft
3. continued Chinese economic growth and consumption of steel, titanium, and other metals that are related to world titanium demand and supply conditions.

Comparing the titanium demand scenarios in Table 6.1 and the supply estimates in Figure 6.1, we observe that the planned expansion of titanium production capacity is largely based on expectations of future titanium demand that most closely resembles the optimistic demand scenario detailed in Table 6.1.

Various combinations of the demand projections and supply scenarios will result in a variety of market situations that are likely to be unbalanced. If producers significantly scale back their capacity expansion plans as strong market demand (similar to the optimistic scenario in Table 6.1) continues, then another supply shortage will result. However, if world titanium sponge capacity is expanded as planned in the coming years, and ingot and mill product capacities match sponge capacity,[12] while demand grows as in the base demand scenario, the result could be a titanium surplus by 2010.

Developments in Titanium Production Technology

In addition to considerations of macro-level supply and demand trends in the titanium market, the future of the titanium industry also could be influenced by technological innovations that help drive down tita-

---

[12] According to industry experts, capital investment for titanium production capacity expansion by titanium sponge producers, ingot melters, and mill product forgers is estimated to be $1.5 billion.

nium production costs. This section will attempt to address the following questions:

- Are there any emerging technologies that may reduce titanium production costs significantly?
- How soon could those technologies be applied to the relevant industry?
- How likely are they to be adopted and commercialized?
- To what extent could they reduce titanium production costs?

To evaluate the potential of various emerging technologies, we met with experts on each phase of titanium production that is illustrated in Figure 6.3. Based on these expert interviews, as well as our own literature review, we developed a list of technologies that had the most potential for titanium production cost savings.[13] Below is a summary of our findings based on our interactions with the experts and our understanding of the industry and existing literature.

Emerging Production Techniques

As discussed in Chapter Two, titanium production involves several phases. Ore is refined into sponge, sponge is melted into ingot, ingot is processed into mill product, and mill product is fabricated into parts. Emerging technologies in each production stage are listed in Figure 6.3. Technologies in roman type already exist; those in boldface italics are emerging technologies. A solid line represents an established production process, while a dashed line is a potential production process. We discuss each of these emerging technologies and their cost-saving potentials below.

[13] Typical questions posed to the industrial technology experts are summarized in Appendix B, Section II. Given our limited time and resources, we could not interview a large number of individuals. Out of 17 individuals we contacted, nine responded to our interviews. Respondents include four experts from the titanium user industry, three from the titanium manufacturing industry, and two from a research laboratory and consulting company.

Figure 6.3
Emerging Technologies of Titanium Production

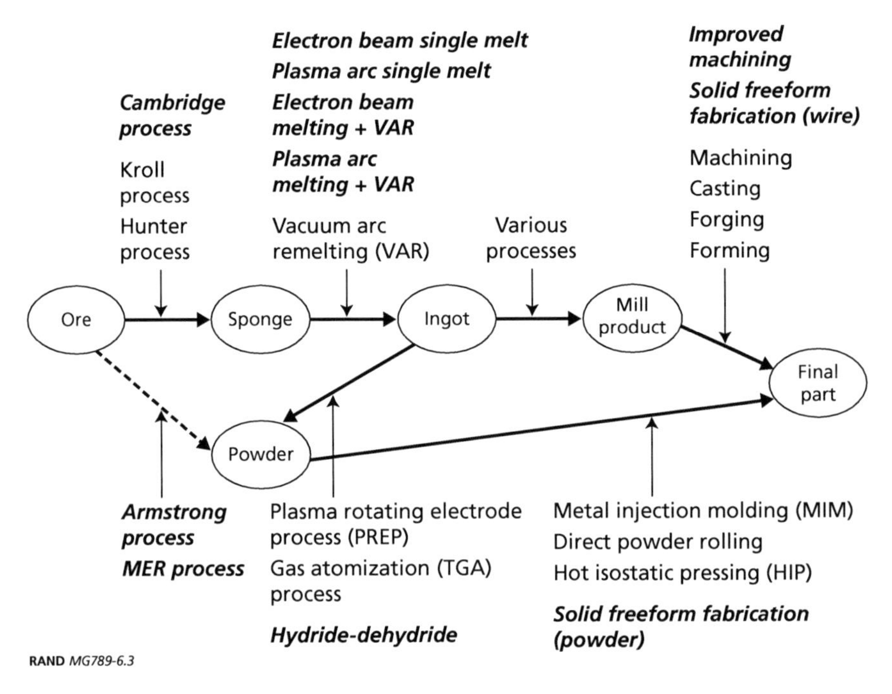

RAND *MG789-6.3*

Improved Titanium Extraction and Refinement

Technical Background. Many researchers are seeking options to replace the Kroll process with a faster alternative that requires less labor and energy. The Cambridge process, under development by FFC Metalysis, uses electrolysis to continually produce titanium sponge. The Armstrong process, developed by International Titanium Powder, produces titanium powder by continuous, low-temperature reduction of titanium tetrachloride. The MER process, under development by DuPont and MER Corporation, produces titanium powder using direct electrolytic reduction. Finally, the hydride-dehydride process, under development by ADMA Group, uses a modified Kroll process to directly produce titanium powder from scrap, turnings, and other titanium waste products.

Notably, three of these four emerging extraction and refinement processes produce titanium powder.[14] Titanium powder currently accounts for an extremely small proportion of annual titanium production. It is produced by atomizing a mill product feedstock using either the plasma rotating electrode process or the gas atomization process (Titanium Information Group, 2006).[15] Because it can only be produced from refined titanium, titanium powder is extremely expensive and is used only for highly specialized applications, such as medical implants. A method for processing titanium powder directly from ore, as the emerging processes do, would make powder much cheaper and, potentially, a much more important part of the market.

Advantages and Challenges. These emerging processes all would reduce costs by increasing production, cutting energy use, requiring less capital, and in most cases using continuous production to reduce the amount of labor needed. The powder production processes do not require VAR, EBM, or PAM to produce a final part, thus removing many steps from the current process and eliminating substantial infrastructure. If successful, their increased speed would expand the amount of titanium on the market and considerably shorten lead times. Savings in energy and labor could also drive down the price of titanium metal, although this depends on the willingness of producers to pass along their savings to users. The primary challenge of these processes is simply getting them to work. The development of each process is discussed below.

Incorporation into the Supply Chain. The Cambridge process produces sponge directly and would be relatively simple to incorporate into current production processes, unlike the other three processes, which produce titanium powder. Powder could be incorporated directly into the existing supply chain as a sponge substitute, but this would negate much of the savings and be commercially impractical due to the high production cost of the titanium powder. However, if the Cambridge

---

[14] The Armstrong, hydride-dehydride, and MER processes produce titanium powder.

[15] The feedstock is a fully processed mill product, such as aerospace grade titanium (Ti-6Al-4V); its composition may vary depending on the final application of the powder.

process were used for powder metallurgy, the savings would be substantially larger. Powder metallurgy is discussed further below.

The success of one of these processes would radically alter a market that has been based on the same technology for decades and dominated by a few companies. Of the four processes identified above, only one, the Cambridge process, was funded by a major sponge producer.[16]

Although new extraction technology could offer a substantial advantage to an existing company or provide the basis for the emergence of a new force in the industry, exploiting this advantage would require a substantial investment in production capacity. Current facilities are optimized for the Kroll process and would need an overhaul to utilize new technology. As such, a new extraction process would be important but would take time to penetrate the market.

Prospects. Although improved refining could yield significant cost savings, most of the technologies mentioned are unlikely to succeed in the near term, if ever. According to industry experts, the Cambridge process has been near commercialization for years, but experts believe it is stymied by still unresolved technical barriers. The Armstrong process has been used to produce small amounts of powder for years, but it has not been scaled up and is not close to large-scale commercial production of powder yet. Some believe the MER process has potential, but it is still early in development and its prospects may not be clear for years.

Hydride-dehydride production is the most promising, and ADMA is reportedly searching for capital to expand its production. Experts believe the process could produce titanium plate for the U.S. Army within several years. However, it could not completely replace the Kroll process, as its low cost comes from higher utilization of titanium scrap. In summary, technical changes in the titanium extraction and refining process are likely to occur over the long-term and remain technically uncertain at present.

---

[16] TIMET was involved in the development of the Cambridge process until it was sold to FFC Metalysis.

Titanium Powder Metallurgy

Technical Background. Titanium powder metallurgy (P/M) is not a new technology; rather it is an entirely different approach to producing titanium parts. A well-developed technology, it has seen wide use with other metals, such as aluminum, but only limited application with titanium. P/M relies on powdered metal rather than sponge to produce parts.

The powder is processed into parts in several ways. Metal injection molding uses a binder to hold the powder together; the binder is burned off during heating. In direct powder rolling, the powder is rolled into sheets, followed by heating and sintering for consolidation. In the hot isostatic pressing (HIP) casting process, the powder is pressed and heated to produce a near-net-shape casting.[17] All of these processes are well developed and currently used to produce a small numbers of parts.

Advantages and Challenges. P/M limits the waste associated with traditional titanium production. It does not require VAR, EBM, or PAM. The powder can be formed directly into any shape or mill product, giving a much higher yield. The process can also produce near-net-shape parts, cutting the waste traditionally associated with part fabrication. By improving the yield at several steps, P/M reduces the amount of raw metal required to make a final part.

However, P/M does introduce several new problems. The first is contamination control, which has posed a problem in the past and would pose a particular challenge for aerospace metals. Any contamination could make the metal unsuitable for aerospace parts, so processing must take place in a clean-room environment. Additionally, it is difficult to use powder to build the large parts that account for a significant portion of the titanium in an aircraft. Finally, titanium powder is explosive, which could create difficulties when handling large quantities of powder.

Incorporation into the Supply Chain. While P/M is currently a niche market, the emergence of an economical source of powder would make these processes very attractive. The high structural standards of

---

[17] Titanium for higher-stress applications usually undergoes an additional VAR melt.

aerospace design make the appearance of P/M in the aerospace market unlikely for some time. It likely would appear first in armor for military ground vehicles, for which the structural requirements are much less rigorous. Although this would not directly affect the aerospace market, it could reduce prices by lessening demand for sponge.

P/M will most likely enter the aerospace market as a new way to produce mill product, which could be used with existing fabrication techniques. Ultimately, it may be applied to near-net-shape castings and other forms of production, which currently have limited use in airframes. The speed of the introduction of P/M will be most affected by materials certification. The expense and difficulty of the certification process will lead companies to avoid P/M products on an aircraft until they have a significant history in other applications.

Prospects. Currently, the main aerospace applications of P/M are in engine production. It is used to produce superalloys that can tolerate extreme conditions in the hot sections of the engine (Eisen, 1996). Because of the contamination control issues, P/M is typically used when there is no other way to produce the desired alloy. In addition, P/M is used to produce several small parts used in military aircraft.

More widespread use of P/M in aerospace would require a large investment near the beginning of the supply chain by the manufacturers of raw titanium and titanium mill products. Titanium powder is easily contaminated and highly volatile, and developing the knowledge and facilities to handle it safely could take years. Depending on how inexpensively powder can be produced, there could also be significant changes to the industry, as discussed above. If P/M does alter the business of titanium production, it will happen incrementally over many years.

Single-Melt Processing

Technical Background. Titanium processing that requires two or three VAR melts has been standard for more than 50 years. Recent developments have attempted to eliminate multiple meltings, allowing titanium processed with a single EBM or PAM to be used for most aerospace applications.[18]

Advantages. Single-melt processing eliminates one melting stage, improving yields and saving time and energy. It also produces titanium at a higher rate than traditional VAR melting. While double VAR takes more than 20–25 days to produce a 3–7-ton batch of titanium, a single EBM can produce more than 15 tons in less than 14 days.[19] EBM and PAM can also utilize smaller and less-pure pieces of scrap, spurring greater demand for recycled titanium.

These strengths make single-melt very useful for mass production, although it is poorly suited for producing small quantities of specialized alloys. It is also harder to control the chemical composition in EBM and PAM, but this appears to be a minor problem.

Incorporation into the Supply Chain. Single-melt is already used commercially and is unlikely to produce any major changes in the market. Because it would require significant capital investment by the major titanium producers, it has been slow to penetrate the market. However, producers are currently expanding production capacity because of high titanium prices, and single-melt should become much more common for new facilities in the near future.

The major obstacle to wider use is the certification of single-melt materials. Boeing and TIMET have been negotiating a standard for single-melt materials for several years, but an agreement has not yet been reached. Nevertheless, both companies believe they are close to an agreement, indicating that adoption of single-melt is less a question of if than of when. Even when single-melt becomes more widespread, buyers may not see an immediate cost reduction. The tight market allows producers to keep the savings as additional profit, and a market downturn may be necessary before savings are passed along to end-users, including aircraft manufacturers.

Prospects. Over the long term, single-melt should become standard. While it is not yet used by commercial producers, it has been

---

[18] Recall that titanium for higher-stress applications usually undergoes an additional VAR melt.

[19] Authors' discussion with TIMET, July 2007.

used in a number of military programs. Structural parts in the tail of the F-15 have been produced with single-melt material, as have static engine components and engine compressor blades. With a strong and growing history, certification for military programs should be less of an issue. The primary limits will be the rate at which single-melt capacity expands and the degree to which price competition forces the major producers to replace their VAR capacity.

Solid Freeform Fabrication

Technical Background. Solid freeform fabrication uses a point heat source to melt titanium wire in successive layers, building up a complex, three-dimensional part. Alternatively, titanium powder may be melted to form the part. A variety of companies, such as Arcam and AeroMet, have developed this technology, typically using either an electron beam or a laser as the heat source.

Advantages and Challenges. These processes are significantly faster than existing techniques of part fabrication, requiring less energy and producing less waste. Despite these strengths, parts produced by solid freeform fabrication can still be more expensive than traditionally produced parts because of the labor and technology required (MTS, 2005). It is suited mainly to small parts, which constitute a small portion of the material used on an aircraft. Its restricted application limits its cost-saving potential and its impact on the cost of military aircraft.

Prospects. This technique was demonstrated by AeroMet Corporation, which developed the technology in conjunction with Boeing and the Defense Advanced Research Projects Agency. As part of the Boeing contract, AeroMet conducted full-scale ground tests with several F/A-18 wing parts and was able to certify them for production use. Ultimately, AeroMet went out of business in 2005, but expert interviews indicate that parts produced using similar techniques have since flown on military aircraft.[20]

---

[20] Authors' discussion with Boeing, January 2007.

Improvements in Machining

Technical Background and Advantages. Titanium is extremely difficult to machine. Its reactivity limits temperatures at the tool face and, by extension, machining speeds. In addition, the low thermal conductivity compounds this problem. Its hardness also tends to wear down tools quickly. Historically, these factors drove the cost of machining to nearly twice that of the raw materials.

However, in recent years a variety of improvements have significantly reduced machining costs. For example, in the past, due to uncertainty in structural loads, parts were designed with some margin to account for that uncertainty. Today, improvements in computer design allow for better analysis and reduced margins. Similarly, the machining of a part can be analyzed in advance and the design optimized to reduce labor and to increase throughput. Such improvements are likely to continue, and the cost of machining should continue to fall in the near future.

Incorporation into the Market and Prospects. These changes seem to be the result of incremental, evolutionary improvements in machining, not a single technological innovation. These improvements result from changes near the end of the supply chain at companies that fabricate parts. They include both major users of titanium, such as Boeing and Lockheed Martin, and small, subcontracted machine shops. The diversity of this market and incremental nature of the changes should force the rapid adoption of innovations. These innovations should produce a downward trend in the price of fabricating a final part, rather than sudden savings.

During part fabrication, a high BTF ratio is often the source of significant costs. Improving this ratio, like improving the yield rates during extraction, melting, and mill product fabrication, could reduce the amount of material needed to make a part. Open sources contain few details about work on this area, but it appears to be a major concern for aerospace companies. It is unclear what changes may be on the horizon, but ways of improving BTF should be monitored.

Cost-Saving Potential of Emerging Technologies

To gain a better sense of timing, feasibility, and the extent of potential cost savings from the potential production innovations discussed above, we asked industry experts to answer questions below as precisely as they could.

- How soon would emerging technologies be ready to be applied to the relevant industry?
- How likely are those technologies to be adopted and commercialized?
- To what extent will those technologies reduce titanium production costs?

The potential time frame for the employment of each emerging technology was assessed as near term (one to three years), medium term (four to seven years), or long term (eight to ten years). The feasibility of employment and cost-saving potential was assessed as low, medium, or high.

Note that these assessments are for technological application to the general titanium market, not just the aerospace market. Stringent certification requirements and conservative engineering may prevent a technology from being applied to the aerospace market until many years after it is proven in the general market. Also, cost savings are relative to the overall process currently in use. For example, a fabrication process may save money on a type of part, but that part may account for a small fraction of the overall titanium used in an aircraft. The findings from the discussions with the industrial technology experts are summarized in Table 6.2

While the savings from improved extraction alone are limited, improved extraction would make powder metallurgy, which would produce much greater savings, an economically viable method of production. At present, titanium powder is extremely expensive, but emerging production processes have the potential to make it cheaper than sponge. The combination of the two developments—refinement alternatives to the Kroll process and the developments of affordable P/M—has a great cost-saving potential. Other technologies could yield

Table 6.2
Potential of Emerging Cost-Saving Technologies

Category	Technology	Time Frame	Feasibility	Cost Savings
Improved extraction and refinement	Armstrong process	Mid	Low	Low
	Cambridge process	Mid	Low	Low
	MER process	Mid	Mid	Low
	Hydride-dehydride	Near	Mid	Low
Powder metallurgy	HIP casting	Long	High	High
	Mill product P/M	Long	High	Mid
	Near-net-shape P/M	Long	Mid	High
Single-melt processing	Cold hearth melting	Near	High	Low
Solid freeform fabrication	Solid freeform fabrication	Mid	Mid	Mid
Improved machining	Various	Near	High	High

NOTES: This table does not include results of a systematic survey. Findings are based on insights of industry technology experts and our understanding of the literature and the industry.

cost reductions by replacing more expensive processes currently in use. These changes would produce more gradual, incremental savings but are also more likely. Improved machining technology has especially high near-term cost-saving potential.

Summary: Developments in Titanium Production Technology

The greatest potential cost savings lie in the combination of improved extraction processes and P/M. However, this is likely to occur only over the long term and is still technically uncertain. Independent of one another, the potential savings from these technologies is limited. Successful development of both technologies would open an entirely different production sequence for titanium parts, with substantial reductions in lead time and material required per pound of part. The success of this change is contingent on finding a replacement for the Kroll process that produces powder rather than sponge. This may be possible, but efforts to do so have repeatedly snagged and failed over the past 50 years. Should current efforts succeed, the titanium industry could change steadily, but dramatically, over the next decade. This could lead to significant reductions in the cost of titanium over the long term, although the speed of any change will be limited by the willingness of producers to invest in new infrastructure.

Single-melt refining and improved machining should produce smaller and more gradual reductions in the cost of titanium products. The degree to which these savings are passed on to titanium consumers will likely depend on the growth of new industrial users of single-melt products. Some savings will be passed on immediately, but consumers may need to wait until supply expands to fully realize the cost savings. These improvements will be much steadier and more certain than the savings offered by improved extraction and powder metallurgy, but they will also be smaller.

Across these new technologies, most savings will be realized by improving yields as the result of reduced waste during processing and part fabrication. Improved labor efficiency will yield some savings, especially during the fabrication process. Energy savings should be an important, but much smaller, proportion of the savings, primarily concentrated in improvements during initial extraction and melting.

Emerging technologies have the potential to reduce costs enough to open new markets, such as military ground vehicles, but they are unlikely to challenge the dominance of the commercial aerospace market in the near future. It will be a number of years before these technologies influence the cost of aerospace-grade titanium. However, in the long run, cost-saving technological progress may truly bring down titanium prices, making titanium an economically viable substitute for other materials. The ability to use titanium as an input material may induce new product innovations, which may again cause a large growth in titanium demand.

CHAPTER SEVEN

Conclusions and Policy Implications

In the previous chapters, we analyzed the underlying factors that triggered the unprecedented surge in titanium price since 2004 and reviewed future market prospects and emerging technologies. In this chapter, we summarize the findings of our research and draw policy implications. We suggest how DoD might mitigate the economic risks involved in the titanium market and reduce the cost of raw materials used in military airframes.

What Triggered the Recent Titanium Price Surge?

The recent extreme price volatility of titanium resulted from the coincidence of various price drivers on both the supply and demand sides, as summarized in Table C.1 in Appendix C.

On the supply side, prices of titanium sponge and scrap began increasing sharply even before the commercial aircraft order surge in 2005 and 2006. Scrap was in extremely short supply in 2003–2005 due to the low airplane production rate and heightened demand from carbon and stainless steel production. During 2003–2005, prices of ferrotitanium almost doubled and influenced the prices of its substitutes—specifically, titanium scrap and sponge.

The DLA titanium sponge stockpile depletion in 2005 also coincided with the sponge and scrap market shortage. Since the supply of titanium raw materials was already tight in 2004, the additional demand from the record high level of commercial aircraft orders in 2005 and 2006 intensified the raw material shortage.

On the whole, increased demand for titanium exceeded the available supply of scrap and sponge as well as the production capacity for new titanium metal. World titanium sponge production capacity declined 22 percent between 1997 and 2004. In the United States, sponge production capacity dropped 70 percent between 1995 and 2004. Given the fact that expanding production capacity requires a large capital investment and a long lead time, expansion could not be responsive enough to meet the unexpected surge in demand over the short run. Moreover, given the long record of excess capacity in the industry, titanium producers were reluctant to invest in capacity expansion until they were assured that strong demand would be persistent for at least several years.

On the demand side, the sudden increase in commercial aircraft orders in 2005–2006 was an obvious price driver. This increase in aircraft orders coincided with the simultaneous trend toward increasing the amount of titanium required to produce each aircraft. This combination of factors amplified the effect of increased aircraft manufacturing on titanium prices.

In addition, increases in the military and industrial demand for titanium coincided with the surge in demand from the commercial aircraft industry. Titanium price volatility was further exacerbated by the increase in spot transactions in the titanium market during the 2005–2006 period. During this surge, even aircraft manufacturers, which normally rely on long-term contracts for their titanium, had to procure titanium on the spot market because of the supply shortage and long lead time. The strong seller's market meant that titanium prices were subject to the producer's bargaining power. Titanium producers thus had a golden opportunity to recover from the severe downturn (1998–2003) that preceded the demand surge in 2004.

Because the market imbalance was generated by a series of unanticipated events and factors, it induced a spurt of speculative purchasing that amplified price volatility, particularly in the spot market. With increased spot market transactions and speculative demand, prices skyrocketed and spot market prices became extremely volatile.

China's Impact on Titanium Prices

Chinese demand and the expansion of its production capacity of various metals has been a wild card in the titanium industry.[1] Combined with the world economic recovery, China's dramatic growth in steel consumption caused a rapid increase in the price of ferrotitanium, an alloy used in steel production, especially during the extreme titanium scrap shortage of 2003–2005. The ferrotitanium price surge led to increased demand for titanium scrap and sponge because they are close substitutes for ferrotitanium in steel production. This cross-market substitution effect caused a substantial surge in titanium raw material prices, given the sheer size of the steel market compared with the titanium market.

Responding to this shortage and skyrocketing prices, China increased its titanium sponge production capacity dramatically, over 300 percent in the two years between 2005 and 2007. This contributed to the stabilization of sponge spot market prices since 2007. Due to the large size of China's economy and its remarkably fast and sustained growth, China may continue to be a wild card in the future titanium market.

Market Prospects and Emerging Technologies

The Titanium Market in the Near Future

In response to the recent demand surge, many titanium metal producers announced or took steps to increase sponge capacity in the near future. The world's sponge production capacity is expected to be approximately 217,970 tons per year by 2010, almost doubling the capacity of 2005. If the new sponge plants are in full operation as planned, Japan and China will be the top titanium sponge producers in the world, followed by Russia and the United States.

---

[1] China influenced other metals markets, too. As discussed in Chapter Five, prices of both ferrous and nonferrous metals have been volatile since 2003, partly due to China's dramatic increase in consumption.

To depict future market conditions, we examined three different future scenarios of world titanium demand— base, optimistic, and pessimistic. In each scenario, we assumed a given combination of annual average growth rate in titanium demand from the aerospace and industrial market segments and then calculated the projected demand in 2010 compared with the actual 2005 demand. We based our assumptions about the growth rate of each market segment on our understanding of the market and discussions with industry experts.

By comparing the demand scenarios and production capacity expansion plans, we determined that the titanium industry's current capacity expansion plans appear to be based on the future demand expectations inherent in the optimistic scenario. If the base demand scenario is realized and the world titanium production capacity expands as planned, we expect that there will be excess capacity in the titanium market by 2010.

We did not attach probabilities to each of the different future scenarios; rather, we used them to bound predictions for the future. Assumptions regarding three market factors heavily influence the future titanium market outlook:

1. the realization of the capacity expansion plans by titanium suppliers including American and Chinese producers
2. the Boeing 787 build rate and demand from other titanium-intensive emerging aircraft
3. continued Chinese economic growth, consumption of steel, titanium, and other metals that are related to world titanium demand and supply conditions.

As a result of these variables, different combinations of demand and supply scenarios will result in a variety of potential market imbalances or market equilibria.

Titanium Production Cost Drivers

Titanium is expensive to refine, process, and fabricate. In terms of processing cost of materials per cubic inch, titanium is five times more expensive than aluminum to refine and more than ten times as expensive as aluminum to form ingots and fabricate finished products. Of all the titanium processing stages, fabrication costs the most, followed by sponge production.

Refining Cost. Refining the raw ore into titanium metal is a costly, multistep, high-temperature batch process that is energy- and capital-intensive. Due to the high reactivity of titanium, an extraction process similar to that for aluminum has not yet been developed.[2] Refining titanium metal into ingot is a slow, energy-intensive process that requires significant capital.

Fabrication Cost. The two main factors in the cost of titanium fabrication are slow machining processes and the high BTF ratio in titanium part production.

The hardness that makes titanium so desirable also makes it more difficult to machine than traditional aluminum. This presents a challenge akin to machining high-strength steel. However, this challenge is complicated by the high reactivity and low thermal conductivity of titanium. Because titanium is highly reactive, it tends to wear away tools very quickly, especially at higher temperatures. Its low thermal conductivity means that high temperatures can be generated easily in the course of machining. Consequently, titanium must be machined at lower tool speeds, slowing production.

Buy-to-Fly Ratio. Titanium parts have very high BTF ratios, with most parts machined from large, solid pieces of metal. Because the raw material is so expensive, the excessive scrap represents a significant portion of the product's cost. Not all scrap can be reverted or recycled. For instance, machine turnings that are not carefully segregated from other metals are too difficult to clean and reuse. In the fabrication process, a significant portion of a part's cost is often left on the machining room floor.

Emerging Technologies

Breakthroughs. Most experts do not expect that any breakthroughs in titanium metal extraction and processing technologies will be realized within the next ten years to drive titanium prices as low as

---

[2] We discussed trends in titanium processing technologies in Chapter Six.

those for aluminum. In addition, the titanium industry has not identified any particular technology that is worthy of an aggressive investment for a medium-term (three to five years) return. Titanium companies are taking a "wait and see" position on potentially significant technological breakthroughs. Only a few experts were optimistic about dramatic changes in titanium production technology within the next ten years.

Technologies with Cost-Saving Potential. We developed a list of emerging technologies that had at least marginal cost-saving potentials, based on a review of the literature and discussions with industry experts (see Figure 6.3 and Table 6.2). These technologies were classified into five categories:

- improved extraction and refinement
- powder metallurgy
- single-melt processing
- solid freeform fabrication
- improved machining.

We then assessed the emerging technologies based on the potential time frame for each, the feasibility of application to the market, and the cost-saving potential of each.

Prospects for Cost Savings. The greatest potential cost savings lie in the combination of improved extraction processes and P/M. However, this combination of developments is unlikely to occur in the near or mid term and is still technically uncertain.

Single-melt refining and improved machining should produce smaller and more gradual reductions in the cost of titanium products. The degree to which these savings are passed on to titanium consumers will likely depend on the amount of slack in the market. Some savings will be passed on immediately, but consumers may not realize significant savings until demand slows and the supply expands, forcing producers to reduce prices. Although these improvements will be steadier and more certain than the savings offered by improved extraction and P/M, they will also be smaller.

Across these new technologies in the near term, most savings will be realized by improved yields resulting from reduced waste during processing and part fabrication. Improved labor efficiency will yield some savings, especially during the fabrication process. Energy savings should be an important, but much smaller, proportion of the savings, primarily concentrated in improvements during initial extraction and melting.

Emerging technologies have the potential to reduce costs sufficiently to open new markets, such as military ground vehicles. However, it will take a long time for these technologies to substantially influence the cost of aerospace-grade titanium.

Barriers to Adopting New Technologies. A major barrier to the adoption of new technologies in aerospace applications is the required certification of new materials. Aerospace manufacturing standards are typically based either on judgments by a government body, such as the Federal Aviation Administration or the U.S. Air Force, or on standards set by the primary aircraft manufacturer. Within the Air Force, materials and processes must be certified separately for each program. The certification process typically lasts 18 to 24 months and requires an extensive qualification process. In the course of this process, a company must manufacture test articles and validate their properties at its own expense. The cost of this process prevents companies from attempting to certify materials until they are quite certain of their performance and properties. Consequently, a material must be used for several years in other applications before designers will consider it for aerospace uses.

Policy Implications

Given the findings of this study, what policy measures could mitigate the economic risks of titanium price volatility and supply availability and reduce the cost of raw materials needed for military airframe production? In this section, we suggest policy measures in five areas:

- improving contract practices
- monitoring market trends

- reducing BTF ratios
- optimizing scrap recycling
- exploring new technological opportunities.

Long-Term Contracts Are Needed to Mitigate Market Volatility

Long-term agreements constitute one of the best ways to mitigate the effect of unexpected price increases and other market volatility. LTA prices are usually much more stable than spot market prices. A built-in price stabilizer can be incorporated in the LTAs. Supply availability can be better secured through LTAs than through short-term contracts.

DoD could explore taking advantage of its significant bargaining power by consolidating titanium contracts to achieve a lower, DoD-wide price and leveraging the stability of military demand relative to commercial aircraft industry demand.[3] Even though alloys vary across different programs and platforms, military demand for titanium raw materials is sizable and relatively stable.

LTA prices are usually higher than spot market prices in a down cycle, but LTAs will be lower than spot prices in a booming market. Therefore, during a downturn, titanium buyers tend to prefer spot market transactions. In particular, lower-tier titanium purchasers in the DoD contractor chain tend to rely on annual contracting with some spot buying each year. DoD would benefit if it monitored the procurement contract practices of its prime contractors and subcontractors and their supply-chain relationships. A balanced purchasing agreement portfolio would include a combination of LTAs with fixed prices, LTAs with built-in price stabilizers, annual contracts, and spot transactions.[4]

---

[3] Military buyers are often at a disadvantage in individual contract negotiations, mainly due to their relatively small and irregular orders compared with those from the private sector. DoD-wide consolidation of contracts could partially overcome this disadvantage. DoD has been trying to apply contract consolidation when possible through its Strategic Sourcing Initiative. This is also a policy of the Office of the Secretary of Defense. Services such as the Army and Air Force have commodity councils to implement the policy. So far, we have not found contract consolidation for raw materials. However, raw materials, especially specialty metals including titanium, get more attention from defense buyers and commodity councils because of their great price volatility.

[4] A balanced portfolio of purchase agreements of varying maturities and conditions needs to be enforced in collaboration with private counterparts. Past experience in the titanium

To the extent that DoD values supply availability and price stability, LTAs should still continue to satisfy most of the raw material requirements—even given this diversified approach to purchasing.

To enable LTAs and contract consolidation, the DoD should seek to increase the predictability and consistency of its military aircraft procurement plans and purchases. Aircraft procurement lots often vary greatly during the course of the acquisition process. Not every military aircraft procurement is based on multiyear contracts. During the recent turmoil in the titanium market, some of the military aircraft contractors were exposed to spot market volatility and supply shortages to a greater extent than were their commercial counterparts because they had to purchase titanium for one lot production at a time.[5]

Monitoring Market Trends to Improve Forecasting Power

To better predict market volatility and enable more favorable contracting arrangements, it would be worthwhile for DoD organizations that procure titanium-intensive weapon systems to monitor the major determinants of world market trends.

The size of the titanium market is very small—only one ten-thousandth the size of the steel market and one two-hundredth the size of the aluminum market in terms of annual shipment tonnage. Therefore, relatively small shocks in the supply and demand of titanium or external shocks in other metals markets cause sizable turbulence in the titanium market. A supply-demand imbalance could be worsened significantly by the cyclical nature of the commercial aircraft manufacturing industry, which accounts for about one-half of the global titanium demand.

Like other raw material markets, the titanium market is globalized, and domestic titanium prices fluctuate with those in the rest of the world. Therefore, titanium price and availability forecasts should be

market has shown that leaving the suppliers alone to figure out their contract formulations has not always been in the best interest of DoD, because suppliers sometimes cannot see beyond their own interests.

[5] Multiyear contracts would promote LTAs for raw material procurement. Younossi et al., 2007, illustrates merits of multiyear contracts for F-22 procurement.

inferred from global consumption and shipment patterns rather than U.S. demand alone, which is dominated by the aerospace sector.[6]

Due to the difference in market focus, U.S. titanium mill production trends can be quite different from those in the rest of the world. As we observed from Figure 5.9, world titanium shipments had fully recovered to their previous peak level by 2004 due to the fast growth in the industrial segment of the global market. However, shipments by U.S. producers in 2004 were still only two-thirds of their previous peak level, and U.S. titanium producers were laying off some employees in some of their factories even during the summer of 2004. This may be one of the reasons that U.S. producers were not ready for the market upturn until 2006, a few years after the market's dramatic turnaround.

Demand-side factors DoD should monitor include the following:

- downstream industry conditions for aircraft manufacturers that may trigger cyclical fluctuations in the titanium industry, including
  - forecasts of aircraft orders and deliveries
  - utilization of titanium in new aircraft designs
  - air traffic and passenger-mile growth
- military programs for jet fighters and other titanium-intensive weapon systems around the world
- global economic conditions that could influence the industrial sector's demand for titanium
  - infrastructure growth in developing areas
- new markets for titanium that may provide opportunities for the industry to grow
  - growing demand for oil and gas production from deep-water sources
  - other fast-growing markets for titanium, such as automotive, heavy vehicles, geothermal facilities, and medical devices

[6] Refer to Table 3.1, Figure 3.3, and Figure 3.4 for a comparison of titanium shipments by sector in the United States and in the world.

- market environments and technology trends that may dramatically change prices and shape the supply availability of materials that can substitute for or complement titanium:
  - ferrotitanium
  - vanadium
  - steel
  - other materials that have corrosion resistance, high strength-to-weight ratios, low modulus of elasticity, and shock resistance properties
- other potential external shocks on the demand side, such as titanium supply and demand trends in rapidly growing large countries, particularly China.

Supply-side factors to monitor include the following:

- global production capacity of titanium sponge, ingot, and mill products
  - capital investment plans and executions
  - new entries and exits in the titanium industry
- changes in the pricing system that may influence titanium price volatility, such as the significance of spot market transactions and contractual practices between major buyers and suppliers
- technical changes in titanium metal production that may reduce not only production costs but also the cost of investing in production facilities
- changes in manufacturing techniques and processes for titanium aircraft parts that may significantly reduce the cost of parts
- trends in the inventory of titanium scrap.

Reducing BTF Ratio and Optimizing Scrap Recycling

High BTF ratios for titanium parts are one of the most significant production cost drivers. It would be helpful for each military aircraft program to exploit the possibility of reducing BTF in every manufacturing stage. With the advent of three-dimensional solid design techniques, parts can be designed such that material can be used more efficiently and minimize waste. Ultrasonic modeling and inspection enable near-net-shaped forgings, reducing machining processes and material usage.

Given current aircraft parts manufacturing technologies, it is inevitable that some scrap will be generated. Efforts should be made to improve the scrap management and recycling rates for machine turnings and bulk solids so that less material is wasted. The private sector uses various strategies to maximize the benefit of scrap recycling, especially since the recent titanium price surge.

Scrap recycling is often embedded in contractual agreements for titanium transactions as a discount factor. This type of scrap recycling is called a "closed-loop arrangement" and is preferred by both titanium producers and consumers. The military sector should leverage these market best practices from the private sector to optimize material usage. In addition, DoD may benefit by using titanium from retired aircraft such as the F-15 and F-14 fighters. There is a significant amount of structural titanium in these aircraft that could be used as scrap and reprocessed into ingot for future aircraft production.

Searching for New Technological Opportunities

Developing or finding new alloys and substitutes is another policy alternative to reduce the economic risks of the titanium market. DoD may benefit from its continued efforts to find alternative materials, such as bismaleimide, ceramic, carbon, and metal-matrix composites. In some cases, these composites can serve as a close substitute for titanium because of their light weight and endurance under high temperatures.

As we discussed above, emerging technologies for titanium production and processing are expected to generate significant cost savings only over the long term. Given that the private sector is not always aggressive in pursuing long-term research and development projects, DoD may want to allocate some of its research capabilities or resources to developing cost-effective methods of extracting, producing, and fabricating titanium. DoD could also be a catalyst to encourage industry collaboration and provide incentives for the private sector to pursue joint ventures in these areas. These efforts could generate long-term benefits not only for DoD but also for the global titanium market and the economy in general.

APPENDIX A

Aircraft Included in the Titanium Demand Calculation and Data Sources

Table A.1
Aircraft Included in the Titanium Demand Calculation and Data Sources

Aircraft Included in Titanium Demand Calculation			Data Sources	
Type	Manufacturer	Aircraft Name	Number of Planes Delivered Each Year (Historical Data)	Titanium Buy Weight per Each Aircraft
Commercial	Boeing	717	Boeing Web site	Holz, 2006
		737		TIMET, 2005
		747		TIMET, 2005
		757		RMI Titanium Company, 1994
		767		Holz, 2006
		777		TIMET, 2005
		787		TIMET, 2005
	Airbus	A300	Airbus Web site	Holz, 2006
		A310		ASM International, 2000
		A318		Holz, 2006
		A319		Holz, 2006
		A320		TIMET, 2005

Table A.1—Continued

Aircraft Included in Titanium Demand Calculation			Data Sources	
Type	Manufacturer	Aircraft Name	Number of Planes Delivered Each Year (Historical Data)	Titanium Buy Weight per Each Aircraft
		A321		Holz, 2006
		A330		TIMET, 2005
		A340		TIMET, 2005
		A350		TIMET, 2005
		A380		TIMET, 2005
Military	Boeing	C-17	World military and civil aircraft briefing	ODUSD-IP, 2005
		F-18 C/D		RMI Titanium Company, 1994
		F-18 E/F		ODUSD-IP, 2005
		F-15		Schmitt, 1993
	Lockheed Martin	F-16		RMI Titanium Company, 1994
		F-22		F-22 Program Executive Office
		F-35		F-35 Joint Program Office, Lockheed Martin Aeronautics Company, ODUSD-IP, 2005

NOTES: The data sources of the aircraft deliveries in the fourth column are for historical data. Forecasted commercial aircraft delivery data are from *The Airline Monitor.* Titanium buy weight per aircraft includes both the airframes and engines.

APPENDIX B

Questionnaire to Industry Experts

I. Titanium Raw Materials Market

I.1 Contract Practices and Titanium Pricing

- Do aircraft manufacturers secure their titanium supply at the time of aircraft orders?
- What is the average length of contracts with aircraft manufacturers?
- What are the main determinants of the price? Is the price contingent on time of delivery?
- If it is, what is the time lag between the delivery of titanium materials and delivery of the aircraft?
- Is the shipment schedule to the aircraft manufacturers usually coordinated with their production schedule?
- Are the sales statistics of titanium producers based on shipment each year or the contract amount (including the long-term contracts) for the year?

I.2 Recent Titanium Price Surge

- Industry experts seem to agree that the recent titanium price surge was mainly driven by demand-side factors such as aircraft orders and strong demand from industrial sector and emerging markets. Do you agree with this view?

- How was the market condition before 2004?
- It seems that titanium mill product PPI did not decrease during the period of low aircraft orders (1991–1994, 2001–2003). Why was the PPI not responsive to decreased aircraft orders?
- Is this related to long-term contract practices and the pricing rules in the contracts? Or is this related to increasing demand from the industrial sector?
- Some industry experts emphasize the shortage of sponge and scrap triggered the recent titanium mill product shortage and price surge. Do you agree with this view?
- How serious was the shortage of titanium sponge and scrap?
- Did the specialty metals clause (Berry Amendment) influence in any way the domestic titanium metal shortage and surge in prices?
- How did the depletion of the national stockpile of titanium sponge influence the titanium metal prices in recent years?
- Was there any hoarding (hedge buying) of titanium sponge, ingot, and mill products for the purpose of securing future supplies?

I.3 Impact of Titanium Price Surge

- Was the impact of the sponge and scrap shortage distributed evenly to all end users? Were industrial users influenced by the shortage and price surge more than aerospace users? If so, why?
- Do the long-term contract practices of the aerospace industry help the aerospace sector stay less vulnerable to the shortage of titanium and price surge?
- Do industrial users of titanium depend on spot markets more than long-term contracts?

I.4 Titanium Supply Elasticity

- How responsive is the supply of titanium metal to price fluctuations?
- What makes titanium producers expand production capacities? Would persistent increase in price be a trigger for capacity expansion?
- Please share with us the past history and future prospects of titanium production capacity (sponge, ingot, and mill products) and its responsiveness to price.
- It seems that the nominal unit price of sponge had been quite stable (price of sponge in real terms had been decreasing) before 2004. Was there worldwide sponge overcapacity since the end of the cold war?
- Data show that U.S. producers had been decreasing sponge capacity, while increasing ingot capacity, for a decade before the recent demand surge (1995–2004). Why? It seems that the ingot capacity utilization rate has been pretty low until recently. Is some of the ingot capacity outdated but still there for statistics?
- Titanium producers started taking steps to expand sponge capacity. Would this expansion be enough to meet the increasing demand for the next several years?
- Should ingot and mill products production capacity be expanded to meet the increasing titanium demand for the next several years?

I.5 Titanium Yield Rate

- Are there any data on how much sponge is needed to produce one metric ton of ingot and how much ingot is needed to produce one ton of mill product?
- For example, we gather that one pound of ingot uses 0.54 pounds of sponge and 0.46 pounds of scrap. One pound of ingot will yield 0.62 pounds of mill product and 0.16 pounds of final parts on average. Does this sound right to you?

- Does ingot for industrial use use relatively more scrap than average?
- Does ingot for aerospace use use relatively more sponge than scrap?
- What, then, are the different yield rates for different end markets?

I.6 Prospect of Cost-Savings in Titanium Metal Production and Fabrication

- Do you envision a significant cost reduction in titanium production and fabrication in the near future?
- To what extent will this cost saving influence the titanium price?
- How soon would this happen?

I.7 Role of the Spot Market

- Is the recent shortage of titanium supply and the price surge mainly a spot market story?
- In the titanium market (sponge, ingot, and mill products) in general, how significant are spot market transactions relative to long-term contract transactions?
- Do aircraft manufacturers depend not only on long-term contracts but also on the spot market to procure titanium raw materials?
- For aerospace buyers, how significant are spot market transactions relative to long-term contract transactions?
- How do you envision the future of the titanium spot market? Will spot market transactions be the dominant form of transactions in the titanium market in the future? How soon will this happen?

I.8 Supply Chain

- In the aerospace sector, are titanium raw materials purchased directly by aircraft manufacturers (Boeing, Airbus, etc.)?

- Would you please describe the structure of your titanium parts supply chain by aircraft name, supplier name, supplier status (prime, second-tier, and third-tier subcontractors, etc.), and their role in the production process?
- Are titanium raw materials mainly procured by tier-1 prime contractors for parts manufacturing? If this is the case, do the part suppliers depend on both the spot market and long-term contracts for securing titanium supply? What would be the relative importance of spot market transactions compared with long-term contract transactions for the prime contractors?
- Is titanium procurement often done by tier-2 and tier-3 parts suppliers (subcontractors of the prime contractors)? Do these lower-tier suppliers tend to depend more on the spot market?

I.9 Other

- What could the government (or DoD) do to reduce the price volatility in the titanium market?
- What were the direct and indirect impacts of Chinese consumption of raw materials on the titanium market in recent years?

II. Titanium Parts Manufacturing Technologies and Cost Drivers

II.1 What are the different steps (machining, welding/joining, etc.) to produce aircraft parts from titanium mill products? Which of these steps are the main cost drivers?

II.2 What are the buy-to-fly ratios for different manufacturing processes (machined, HIP, CIP, forged, extruded, etc.)? Which of these processes are considered "near-net-shape"? What proportion of parts is produced using each of those processes (in terms of number of parts, input titanium weight, or final titanium part weight)? How do you expect these proportions to change in the future?

II.3 What types of parts are currently produced using HIP/CIP? How do these processes yield cost savings (reduced raw material, reduced labor, etc.) and how significant are these savings? What are the drawbacks of this technology that could limit its future applications?

II.4 Do you anticipate making use of powder metallurgy to produce aerospace parts? If so, what types of parts would be produced and when would it be possible to produce them? Would powder metallurgy yield significant cost savings? Would the application of powder metallurgy to non-aerospace uses impact production costs for aerospace parts made from regular titanium?

II.5 Is high-speed machining (HSM) currently applied to the production of aerospace parts? If so, what types of parts employ this technology and what proportion of parts (by number of parts, input titanium weight, or final titanium weight) is produced by HSM? How much does HSM save and by what mechanisms (reduced labor, faster production rates, etc.)? Does HSM have the potential to produce increased savings in the future? If so, how significant will these savings be?

II.6 What parts can currently be made from single-melt titanium? What is the difference in cost between single-melt and double-melt titanium parts? What proportion of the total titanium usage by weight is single-melt? Do you expect the proportion of single-melt use to increase over the next ten years?

II.7 Do you expect solid freeform fabrication (e.g., laser forming) to be used to produce parts within the next ten years? If so, what types of parts? How will this technology produce savings (reduced labor, reduced waste, etc.) and how significant will these savings be? How extensive will use of this technology be in ten years (i.e., how many pounds of titanium parts could it produce per year)? Will production capacity be enough to significantly reduce the costs of using titanium in an aircraft? Will this technology be used in the production of military and commercial aircraft, and, if so, will its use differ in these two areas? Does this technology have significant disadvantages?

II.8 Are there any other technologies not discussed above (i.e., not HIP/near-net-shape, HSM, powder metallurgy, single-melt, solid freeform fabrication) under development that are likely to yield significant cost savings in the production of titanium parts in the next ten years?

II.9 Will the 787 affect titanium material supply for military aircraft, and if so, how? Are the titanium parts manufacturing technologies used in 787 production likely to find their way into production of military aircraft? How long will this take?

II.10 Below is a summary table of titanium production technologies.

- If we missed or misunderstood anything, please revise the first four columns (technology, purpose, remaining challenges, and technologies) as you like,
- Please populate the last three columns.

In the "cost-saving potential" column, you may put down low, medium, or high according to the extent of cost-saving potentials of each technology.

In the "feasibility" column, you may put down the likelihood that each technology will deliver enough cost savings to justify its use and be at a level of development where it can be applied to a production-model military aircraft—low, medium, or high.

In the "time frame" column, you may estimate when the technology will be developed and ready for use in a program—near (1–3 years), medium (4–7 years), far (8–10 years).

Table B.1
Emerging Technologies and Their Cost-Saving Potentials

Category	Purpose	Remaining Challenges	Technology	Time Frame	Feasibility	Cost Savings
Improved extraction and refinement	Reduce materials cost	No history. Most processes still in development	Armstrong process Cambridge process MER process Hydride-dehydride			
Powder metallurgy	Easier alloying eliminates many processing steps. Near-net-shape production	May not scale easily. Minimal history	HIP casting Mill product P/M Near-net-shape P/M			
Single-melt processing	Simplifies refining	May produce lower quality	Cold hearth melting			
Solid free-form fabrication	Near-net-shape production	Only at demonstration phase	Solid free-form fabrication			
Improved machining			Various			

APPENDIX C

Supply- and Demand-Side Conditions Resulting in the Recent Titanium Market Turmoil

Table C.1
Titanium Supply- and Demand-Side Events, Early 1990s–2006

Time Line	Demand-Side Events	Supply-Side Events
Early 1990s	Titanium demand decreased since the cold war	Excess capacity since the end of the cold war
Mid 1990s	Industrial demand for titanium surpasses aerospace demand for titanium in the world market	Number of U.S. titanium sponge and ingot producers starts to decrease (from 11 in 1995 to 5 in 2005) No new entrant in titanium sponge and ingot manufacturing industry from mid 1990 until 2007
1997	Titanium demand from commercial aircraft manufacturing peaks in 1997 Titanium mill product shipments peaked in 1997 Titanium demand from commercial aircraft industry starts declining after 1997; downswing continues to 2003 World titanium shipments show downward trend between 1997 and 2002	Congress authorizes disposal of the DLA titanium sponge stockpile Until 1997, titanium sponge stockpile was maintained at about 33,000 tons to cover total domestic consumption for at least one year, even during peak consumption
1999	Initial production of the F-22 starts	
2001	Industrial demand for titanium starts to swing upward	

Table C.1—Continued

Time Line	Demand-Side Events	Supply-Side Events
2002	Industrial demand for titanium starts to grow	
2003	Aircraft manufacturing bottoms out World titanium mill product shipments are about 83% of the 1997 peak U.S. titanium mill product shipments are only 56% of the 1997 peak Full-time production of the F-22 begins and military aircraft demand for titanium increases significantly	Scrap supply is extremely low as aircraft manufacturing reaches bottom Scrap demand peaks as growing carbon steel and stainless production caused ferrotitanium demand to surge Ferrotitanium price almost doubles in 2003–2005 Titanium raw materials (scrap and sponge) prices increase sharply in 2003–2005 due to the cross-market substitution effect from steel industry DLA titanium sponge stockpile is about 1/5 that of 1997, and titanium producers continuously use the sponge stockpile as a substitute for titanium scrap and ferrotitanium
2004	World economic growth peaks, driven by growth in China, the Middle East, and other developing areas World steel production peaks World titanium mill product shipments reach their previous peak level of 1997 U.S. titanium mill product shipment are still only 68% of their previous peak level, as aircraft manufacturing industry recovery has not yet started	Titanium scrap and sponge prices increase 71% and 49%, respectively World titanium sponge capacity in is 22% lower than the peak capacity in 1997 U.S. titanium sponge capacity in is 70% lower than the 1997 capacity U.S. titanium producers hesitate to invest in capacity expansion, having suffered several years of bad business

Table C.1—Continued

Time Line	Demand-Side Events	Supply-Side Events
2005	Historic level of commercial aircraft orders in 2005 and 2006, which were not expected ahead of time Commercial aircraft orders for Boeing and Airbus are 2,139 aircraft in 2005 and 1,882 aircraft in 2006, twice as many as in the previous peak years Due to the unexpected nature of the demand surge, spot market transactions increase Speculative buying and hoarding occur as supply shortage becomes serious	Prices of steel products stabilize as China turns into a net exporter of steel Titanium sponge stockpile of DLA finally depleted in 2005 In the strong seller's market, titanium prices subject to the bargaining power of suppliers World titanium sponge capacity grows 10%; U.S. capacity does not increase
2006	U.S. titanium mill product shipments recover fully to their previous peak level of 1997 Titanium spot market prices especially volatile: Titanium ingot used in aerospace manufacturing quadrupled in the three years from 2003 to 2006 Titanium delivery lead time increases by up to three times	Scrap shortage situation improves as aircraft production increases significantly and generates scrap World titanium sponge capacity grows 15% US.. sponge production capacity increases 38%

Bibliography

Aboulafia, Richard L., "World Military & Civil Aircraft Briefing," Fairfax, Va.: Teal Group Corporation, January 2007.

Aerospace Industries Association of America (AIA) Web site. As of September 27, 2008:
http://www.aia-aerospace.org/

Airbus, Airbus for Analysts, Web site. As of March 21, 2008:
http://www.airbus.com/en/airbusfor/analysts

The Airline Monitor, "A Review of Trends in the Airline and Commerical Jet Aircraft Industries," 2005. As of March 21, 2008:
http://www.airlinemonitor.com/

———, "Update of Commercial Aircraft Market Forecast, Financing Aircraft Deliveries 2007 to 2020 and Update of Engine Deliveries 2007 to 2025," January–February 2007.

American Institute of Aeronautics and Astronautics (AIAA) Web site. As of September 27, 2008:
http://www.aiaa.org/

American Iron & Steel Institute (AISI) Web site. As of March 21, 2008:
http://www.steel.org//AM/Template.cfm?Section=Home

ASM International, *Titanium: A Technical Guide*, 2nd Edition, Bilthoven, The Netherlands: ASM International, 2000.

Barksdale, Jelks, "Titanium," *The Encyclopedia of the Chemical Elements*, Skokie, Ill.: Reinhold Book Corporation, 1968, pp. 732–738.

Berry Amendment Reform Coalition, "Senate Berry Amendment Streamlining Proposal: Myth Versus Reality," July 18, 2006. As of September 16, 2008:
http://www.aia-aerospace.org/pdf/legbrief_berrymyths_071806.pdf

Boeing, Commercial Airplanes: Orders and Deliveries, Web-based data search tool. As of March 21, 2008:
http://active.boeing.com/commercial/orders/index.cfm?content=userdefinedselection.cfm&pageid=m15521

Bureau of Labor Statistics (BLS), Producer Price Index Statistics. As of March 21, 2008: http://www.bls.gov/data/home.htm

Bush, Jason, "Boeing's Plan to Land Aeroflot," *Business Week*, February 15, 2006. As of March 21, 2008:
http://www.businessweek.com/technology/content/feb2006/tc20060215_694672.htm?campaign_id=search

Cariola, Monica, "A High-Potential Sector: Titanium Metal Oligopolistic Policies and Technological Constraints As Main Limits to Its Development," *Resources Policy,* Vol. 15, 1999, pp. 151–159.

Chierichella, Hon W., and David S. Gallacher, "Specialty Metals and the Berry Amendment: Frankenstein's Monster and Bad Domestic Policy," *The Government Contractor,* Vol. 46, No. 16, April 2004.

Churchill, David A., and Kathy C. Weinberg, "Domestic Specialty Metals Restrictions: A Bumper Crop of Fresh Berry Issues," Briefing Papers, Thomson West, March 2007.

Defense Acquisition Management Information Retrieval Web site, Selected Acquisition Report (SAR): F-22A, RCS: DD-A&T (Q&A) 823-265, December 31, 2006.

———, Selected Acquisition Report (SAR): F-35 Joint Strike Fighter, RCS: DD-A&T (Q&A) 823-198, December 31, 2006.

Defense Industry Daily, "Domestic Titanium Requirements Become an Issue in the U.S.," March 20, 2006. As of March 21, 2008:
http://www.defenseindustrydaily.com/2006/03/domestic-titanium-requirements-become-an-issue-in-us/index.php

Department of Defense (DoD), *Report to Congress: An Integrated Plan for the Development and Processing of Low-Cost Titanium Materials and Associated Manufacturing Processes*, August 2004.

———, *Strategic and Critical Materials Report to the Congress*, 2006.

DoD—*See* Department of Defense.

Eisen, William, "Powder Metallurgy Superalloys," *Materials World*, Vol. 4, 1996, pp. 22–24. As of March 21, 2008:
http://www.azom.com/Details.asp?ArticleID=152

Encyclopedia Britannica Online. As of September 27, 2008:
http://www.britannica.com/

Fanning, John, "Titanium Alloys for Composite-Intensive Airframes," TIMET Presentation at IT Conference on Titanium, San Diego, October 3, 2006.

Fifth Supplemental DoD Appropriations Act of 1941, Public Law No. 77-29, 55 Stat.123.

Gambogi, Joseph, "Metal Prices in the United States Through 1998: Titanium," Washington, D.C.: U.S. Geological Survey, 1998.

———, "Titanium," *United States Geological Survey Minerals Yearbook: Volume I: Metals and Minerals*, 2004. As of March 21, 2008: http://minerals.usgs.gov/minerals/pubs/myb.html

Gerdemann, Steven J., "Titanium Process Technologies," *Advanced Materials and Processes*, July 2001. As of March 21, 2008: http://www.itponline.com/index_files/ASMarticle.pdf

Grasso, Valerie Bailey, *The Berry Amendment: Requiring Defense Procurement to Come from Domestics Sources*, Congressional Research Service (CRS) Report for Congress, April 21, 2005.

Haflich, Frank, "Titanium Sponge Prices, Producer's Backlogs Up," *American Metal Market*, February 10, 2005a.

———, "DNSC Sells Last of Titanium Sponge to Three Companies," *American Metal Market*, December 2, 2005b.

Holz, Markus, "European Titanium Market: Current and Future Scenario," Titanium 2006 Conference, San Diego, California, 2006. As of March 21, 2008: http://www.deutschetitan.com/documents/ETM.pdf

Hubbard, Glenn R., and Robert J. Weiner, "Nominal Contracting and Price Flexibility in Product Markets," NBER Working Paper No. 1738, Cambridge, Mass.: National Bureau of Economic Research, 1985.

Hurless, Brian E., and F. H. Froes, "Lowering the Cost of Titanium," *The AMPTIAC Quarterly*, Vol. 6, No. 2, 2002.

International Iron and Steel Institute (IISI), undated. As of September 17, 2008: http://www.worldsteel.org/index.php

International Stainless Steel Forum (ISSF). As of March 21, 2008: http://www.worldstainless.org/

International Titanium Association (ITA), "Titanium: The Ultimate Choice," Broomfield, Colo.: ITA, 2005a.

———, *Statistical Review 2001–2005,* Broomfield, Colo.: ITA, 2005b.

Iron and Steel Statistics Bureau, *World Steel Review,* February 2006. As of March 21, 2008: http://www.steelonthenet.com/ISSB/Review-02-06.pdf

Kraft, Edwin H., *Summary of Emerging Titanium Cost Reduction Technologies: A Study Performed for U.S. Department of Energy and Oak Ridge National Laboratory*, Vancouver, Wash.: EHK Technologies, January 2004.

London Metal Exchange Web site, 2008. As of October 29, 2008:
http://www.lme.co.uk/

Martin, Rick, and Daniel Evans, "Reducing Costs in Aircraft: The Metals Affordability Initiative Consortium," *JOM*, Vol. 52, No. 3, March 2000, pp. 24–28.

Masson, Francis G., "Structure and Performance in the Titanium Industry," *The Journal of Industrial Economics*, Vol. 3, No. 3, 1955, pp. 222–240.

Metalprices.com. As of March 21, 2008:
http://www.metalprices.com/

MTS Systems Corporation, Annual Report to Shareholders 2000, Eden Prairie, Minn., 2000.

———, "MTS Reports Fiscal 2005 EPS of $1.81, up 34 Percent on Revenue Increase of 11 Percent: Company Exits Non-Strategic Businesses," News Release, November 21, 2005.

Murphy, James, "Titanium Price Increases and Shortages May Affect JSF Programme," Janes.com Military Aerospace News, February 28, 2006.

National Defense Authorization Act for Fiscal Year 1998, Section 3304, "Disposal of Titanium Sponge in National Defense Stockpile," Public Law 105-85, 105th Congress, 1997 (House Report 105-304). As of March 21, 2008:
http://thomas.loc.gov/cgi-bin/cpquery/?&sid=cp105MkEXs&refer=&r_n=hr340.105&db_id=105&item=&sel=TOC_1589517&

National Defense Authorization Act for Fiscal Year 2007, Public Law No. 109-364, 109th Congress, 2006. As of Septmmber 25, 2008:
http://www.dod.mil/dodgc/olc/prior_ndaa.html

ODUSD-IP—*See* Office of the Deputy Under Secretary of Defense for Industrial Policy.

Office of the Deputy Under Secretary of Defense for Industrial Policy (ODUSD-IP), "China's Impact on Metals Prices in Defense Aerospace," December 2005. As of March 21, 2008:
http://www.acq.osd.mil/ip/docs/china_impact_metal_study_12-2005.pdf

Phelps, Hank, "Maintaining Material Properties: F-22 Prospective," Presentation of Lockheed Martin Aeronautics-Marietta, GA at the International Titanium Association (ITA) Titanium Conference 2006, San Diego, California, October 1–3, 2006.

RMI Titanium Company, S-2/A SEC filing, June 6, 1994. As of March 21, 2008:
http://www.secinfo.com/dsvrv.b34.htm

RTI International Metals, Inc., "Company Plans to Construct New Premium-Grade Sponge Plant," Press Release, September 17, 2007a. As of March 21, 2008:
http://www.rti-intl.com/

———, "RTI Announces Additional Long-Term Airbus Contract Valued in Excess of $1.1 Billion over Eleven-Year Term," Press Release, September 17, 2007b. As of March 21, 2008:
http://www.rti-intl.com/

Rupert, Timothy G. (President and CEO, RTI International Metals, Inc.), remark at ITA Annual Conference on Titanium, San Diego, California, October 1–3, 2006.

Schmitt, Bill, "Titanium Industry Sets Its Battle Plan; Producers Look to Retain Terrain in Defense Market, as Well as Capture New Patches of Military Demand—Titanium," *American Metal Market*, September 30, 1993.

TIMET *See* Titanium Metals Corporation.

Titanium Information Group, Data Sheet No 18. "Titanium Powder Metallurgy," Titanium Information Group, February 2006.

Titanium Metals Corporation (TIMET), Annual Reports, various years.

Toensmeier, Pat, "Metal Fatigue: Skyrocketing Titanium Prices Raise National Security Concerns," *Aviation Week & Space Technology*, Vol. 164, No. 13, March 27, 2006.

U.S. Geological Survey (USGS), *Historical Statistics for Mineral and Material Commodities in the United States: Titanium Metal,* USGS Web site, 2008. As of March 21, 2008:
http://minerals.usgs.gov/ds/2005/140/

———, Mineral Commodities Summary: Titanium and Titanium Dioxide, various years (survey published in January of each year).

———, Mineral Industry Surveys: Titanium, various quarters and years (quarterly publication).

———, *Minerals Yearbook: Titanium*, various years (annual publication).

U.S. Government Accountability Office (GAO), *Joint Strike Fighter: DoD Plans to Enter Production Before Testing Demonstrates Acceptable Performance*, Report to Congressional Committees, March 2006.

USGS—*See* U.S. Geological Survey.

Younossi, Obaid, Mark V. Arena, Kevin Brancato, John C. Graser, Benjamin Goldsmith, Mark A. Lorell, F. S. Timson, and Jerry M. Sollinger, *F-22A Multi-Year Procurement Program: An Assessment of Cost Savings*, Santa Monica, Calif.: RAND Corporation, MG-664-OSD, 2007. As of September 22, 2008:
http://www.rand.org/pubs/monographs/MG664/

Younossi, Obaid, Michael Kennedy, and John C. Graser, *Military Airframe Costs: The Effects of Advanced Materials and Manufacturing Processes*, Santa Monica, Calif.: RAND Corporation, MR-1370-AF, 2001. As of September 12, 3008: http://www.rand.org/pubs/monograph_reports/MR1370/